WATER IN THE MIDDLE EAST

A
GEOGRAPHY
OF
PEACE

Water
in the Middle East

Edited by Hussein A. Amery
and Aaron T. Wolf

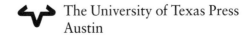 The University of Texas Press
Austin

Requests for permission to reproduce material from this work should
be sent to Permissions, University of Texas Press, P.O. Box 7819,
Austin, TX 78713-7819.

∞ The paper used in this publication meets the minimum
requirements of American National Standard for Information
Sciences—Permanence of Paper for Printed Library Materials,
ANSI Z39.48-1984.

LIBRARY OF CONGRESS CATALOGING-IN-PUBLICATION DATA

Water in the Middle East : a geography of peace / edited by Hussein A.
    Amery and Aaron T. Wolf.
        p.   cm.
    Includes bibliographical references and index.
    ISBN 0-292-70494-1 (cloth : alk. paper). — ISBN 0-292-70495-
x (pbk. : alk. paper)
        1. Water resources development—Political aspects—Middle
East.   2. Water-supply—Political aspects—Middle East.
3. Water rights—Middle East.   4. Jewish-Arab relations.
I. Amery, Hussein A., 1958-   . II. Wolf, Aaron T.
HD1698.M53W383   2000
333.91'00956—dc21   99-30146

Number One
Peter T. Flawn Series in Natural Resource Management
and Conservation

The Peter T. Flawn Series in Natural Resource Management and Conservation is supported by a grant from the National Endowment for the Humanities and by gifts from the following donors:

Jenkins Garrett
Edward H. Harte
Houston H. Harte
Jess T. Hay
Mrs. Lyndon B. Johnson
Bryce and Jonelle Jordan
Ben F. and Margaret Love
Wales H. and Abbie Madden

Sue Brandt McBee
Charles Miller
Beth R. Morian
James L. and Nancy H. Powell
Tom B. Rhodes
Louise Saxon
Edwin R. and Molly Sharpe
Larry E. and Louann Temple

CONTENTS

LIST OF FIGURES AND TABLES

FIGURES

TABLES

Water is an important resource everywhere. It is a particularly important resource in the Middle East and North Africa because the region as a whole entered a phase of water deficit in about 1970. Some countries, the Gulf states, Jordan, Israel, Palestine, and Libya, had run out of water in the 1950s or 1960s. The way the region has handled the transition to water deficit in the past quarter-century and continues to cope in the next few decades is important for the rest of the world. If the MENA region can cope then the pessimists who argue that the world faces a possibly intractable water crisis will have to reconsider their positions.

It is important to specify what one means by running out of water. To provide water for drinking, domestic, municipal, and industrial uses is not generally a difficult challenge. Only about 10 percent of an individual's water consumption is required for these diverse functions, which, together, consume 20 to 100 cubic meters per person per year depending on whether one lives in an industrialized or nonindustrialized political economy. Water to grow the food consumed by an individual needs a much greater volume, estimated at about 1,000 cubic meters per person per year. Much more is demanded in communities that consume high volumes of meat and animal products as a proportion of their diet. In brief, water embedded in food represents about 90 percent of an individual's water needs.

The MENA region could only meet the water needs of its growing population up to 1970 by progressively mobilizing almost all its irrigation potential from surface and groundwater sources. The opportunities to mobilize new water closed for most economies in the early 1970s. Only the Tigris-Euphrates riparians have significant volumes of water for current and future development at relatively low cost. Although high-cost water could also be developed by an expansion of desalination and various water transport options, such water would only be affordable for the domestic, industrial, and service sectors. In agriculture expensive water can only be used to produce a narrow range of high-value crops. The production of staple foods requires inexpensive water.

The transition to serious water deficit in three decades since the early 1970s has been without hydropolitical stress. The explanation lies in theory developed in economic geography. The region has been able to make good its water deficit by "importing" water embedded in staple cereals. It requires 1,000 tons (cubic meters) of water to raise a ton of wheat. It is much easier to move the ton of grain than the 1,000 tons of water. Also the grain on the world market has been heavily subsidized, so water deficit economies enjoy a double benefit when they import such grain. The price of grain has been falling on the world market for the past hundred years.

The chapters which follow are inspired by the interdisciplinary approach of geography. The holistic treatment is essential and well deployed in this book. The study is further enhanced by the in-depth examination of the Jordan basin case. The water resources of concern are properly identified and evaluated, but it is recognized by the authors that hydrology is not enough. Water is an important element in the basin's numerous political economies, and the volume's contributors show that it is usually politics which determines the way water is perceived by peoples and governments. Beliefs and decisions tend to be based on perceptions. The chapters' collective focus on the Jordan basin riparian economies is particularly useful, revealing the ideological inspiration of both the politics and environmental analysis related to these economies. The history of the last half of the twentieth century has been very important in shaping the expectations of the Arab nations with respect to water entitlement and water rights.

That the region has coped with its transition into water deficit can be interpreted in a number of ways. First, within the region, while the political economy of water has been shown to be subordinate to the political economy of the international grain trade, any such awareness has not made the communities and governments of the region feel secure. There is still occasional talk of food self-sufficiency as a goal, although this is becoming less common. But there are still many claims that individual economies, such as Egypt's, are not short of water. Such claims can only mean for domestic and industrial water.

Secondly, some political leaderships have decided to allocate water to staple grain production despite the economic arguments against the use of scarce water in these activities. In the mid-1980s both Saudi Arabia and Egypt swung resources behind the production of wheat. Saudi reversed its policy as a result of the economic impacts of the Gulf War of 1991. Egypt remains committed to its policy of devoting water to cereal production.

Thirdly, outside agencies, inspired by relatively recent experience in the semiarid regions of the United States, have begun to preach the virtues of environmental sustainability and economic water-use efficiency. These principles have only been in scientific currency since the late 1960s. They proved to be fatal to President Jimmy Carter when he tried to introduce them in the United States in his 1976–1979 administration; and they were only introduced just over a decade ago in, for example, California. Policies, based on principles of economic efficiency, which bring a high return to water are politically stressful. Environmentally sound policies often prove costly in the short term. That both these types of policy are being resisted by the communities and politicians of the MENA region is predictable.

The chapters in this volume are useful because they provide a number of very different perspectives on the resource itself, as well as on the region's perceptions of the resource. There are some very well argued contributions which outline and analyze some of the environmental and economic issues discussed above. There are others which examine the evolving international legal principles which could be useful in building alternative and more constructive regimes than those that exist currently. Especially useful are the chapters which emphasize the depth of feeling of political entities which have endured damaging conflict. Approaches forged in such conflictual circumstances confirm the truth that communities use evidence and make decisions on the basis of perception. They certainly base policy on perception in the absence of well-founded and publicly verified information.

The studies reflect well on the editors. They have brought together an eclectic set of chapters which are nevertheless focused on a critical set of issues and united and integrated by the environmental, economic, cultural, and political significance of water in a water-scarce region. The chapters together provide a comprehensive geographical perspective on the "hydrological landscape" of the Jordan basin region. They leave the reader with a clear picture of the hydrological system of the catchment, a "landscape" heavily engineered by the end of the twentieth century. Also provided is an analysis of the consequences of the segmentation of that landscape by contending political entities, for which local water resources are no longer sufficient to meet the water needs of the populations of the basin.

Tony Allan
School of Oriental and African Studies,
University of London

ACKNOWLEDGMENTS

This book was made possible mostly by the hard work of each of the contributors and by their willingness to accept internal and external comments according to which they revised and re-revised their manuscripts. The reader will find a whole spectrum of perspectives and interpretations in this book—compelling and emotional arguments were made by some contributors along the way. The fact that they all stuck it out, despite some tense times, suggests the professionalism of each contributor, and his or her commitment to the project. For that we express out deepest thanks.

We editors also had our share of such arguments. We disagreed sometimes on the interpretation of certain data, and sometimes on the significance of contemporary and historical events. Heated debates were often triggered about the appropriate use of language. We consider our experience as reflective of the intellectual adjustments that are necessary along the rocky road to peace in the Middle East. We certainly have a broader perspective now on the troubles of the region than we did before we started this joint project. But, more importantly, each has learned from, and has greater appreciation for, the perspective of the other.

In addition to the contributors' efforts, the hard work of many others made this project possible. We are extremely grateful to Shannon Davies from the University of Texas Press for her methodical approach, professionalism, and sense of balance at critical junctures. Her keen knowledge of and sensitivity to the issues were evident and appreciated. We would also like to recognize the editorial and intellectual advice of John Kolars, who went above and beyond to round out some rather awkward edges. Thanks, too, to Paul Spragens, whose copyediting expertise kept us all on course. We are also grateful to two anonymous reviewers for their close reading and helpful comments.

Finally, we would like to kindly acknowledge the support, understanding, and patience of our respective families. Our wives, Maha and

Ariella, and our children, Hisham and Yardena, were most understanding and tolerant at times when perhaps they should not have been—when we used to spend entire weekends in the office even on beautiful spring or fall days. Now that the book is going to press, we can both spend more time with our families, and enjoy our newborns, Laila and Eitan.

HAA and ATW
Spring 1998

To our children,
Hisham and Yardena,
Laila and Eitan,
in the hopes that
the day will come soon
when they can picnic
on the banks of the Litani,
then spend the afternoon
boating on the Sea of Galilee

WATER IN THE MIDDLE EAST

1. *Water, Geography, and Peace in the Middle East*
AN INTRODUCTION

Hussein A. Amery and Aaron T. Wolf

BACKGROUND TO THIS BOOK

The idea for this book originated when its editors decided on the need for an edited volume that would have the Jordan River basin as its focal point, a geographical perspective as its analytical approach, and the promise of a Middle East at peace as its planning horizon. It is the inexorable movement toward a Middle East at peace which is the foundation for this work.

We recognize that the peace process proceeds by fits and starts, and has periods of exuberance and mourning. It is just as hard to see the dangers that lie ahead when documents of agreement are being signed as it is to remember the positive possibilities when rocks and bullets fly. Yet just as the Jordan flows unceasingly toward its destination both in times of plenty and times of drought, so, too, we believe the peace process will and must continue to move forward. We must remember how far the peace process has advanced in the short years since negotiations began.

This collaboration between the coeditors, each of whose backgrounds mirrors the other's, was born at a meeting on Middle Eastern water in Waterloo, Ontario, in 1992. Participants, including Israelis and Arabs, had come from around the world. Suspicions were rampant and the rhetoric thick. Each presentation included nine parts of political posturing for each part of water. And yet, a glimmer of hope for peace emerged at that time.

In the relative blink of an eye, given the length of time the two sides have been at odds, attitudes have changed as each side has begun to de-demonize, then listen to, and finally understand the other. Today, Jordan and Israel are developing joint management of the Yarmuk and the Jordan Rivers; Israelis and Palestinians cooperate in joint patrols to police illicit pumping and both sides have recognized the other's water rights.

We come to our task as coeditors from backgrounds representing different sides of the conflict. While our respective conditioning and education in the United States and Canada lead us occasionally to differing interpretations (notice in our chapters, for example, our very different conclusions regarding the relationship between the Litani and Jordan Rivers), we share a fervent belief in the overwhelming potential of a Middle East at peace. We argue, however, that the obstacles in its way not only can be overcome, but will be. Since the people of the region cannot turn back history, they must move forward; they can only do so together.

WATER AND PEOPLE

Nearly three hundred major river basins and as many groundwater resources cross national boundaries. Globally, water availability and quality vary from region to region. The Middle East is of particular concern because of the convergence of volatile factors: conflicting territorial claims, ethnic and historical antagonisms, rapid population expansion through natural growth, immigration and refugee flows, combined with limited surface and subsurface water.

Because political boundaries often ignore this critical resource, and because water flows vary in space and time, disputes are bound to arise, perhaps nowhere more intensely than along the Jordan River. Questions of water ownership and management have exacerbated tensions between Arabs and Israelis, but also have been incentives to dialogue, even when the litigants were legally in a state of war. Currently, all of the riparians of the Jordan basin—Israel, Jordan, Lebanon, Palestine, and Syria— share the common dilemma of inexorably declining per capita shares of water.

The Global Setting

This book, while focusing upon Middle East water, may also be a harbinger of future water negotiations and settlements throughout the world. World water use has increased sixfold between 1900 and 1995 (Commission on Sustainable Development 1997). This is more than double the rate of population growth. The report cited notes that by the year 2025 two-thirds of the world's population will suffer from moderate to severe water stress resulting from overdemand and pollution (see inset on water scarcity). Moreover, the minimal needs of an additional five

THERE ARE FOUR primary causes of water scarcity (Falkenmark and Widstrand 1992; Clarke 1991):

1. aridity and long-term water shortages caused by normal dry
climatic conditions (southern Israel, a small area in northeastern Lebanon, much of eastern Jordan, and southeastern Syria
are thus classified as arid or semiarid);
2. drought, defined as a period in which precipitation is much
lower and evaporation higher than normal (the Jordan River
basin experienced drought starting in the mid-1980s which
lasted until the early 1990s);
3. desiccation, defined as the degradation of the land due to the
drying up of soil as the result of deforestation and overgrazing;
4. water stress, defined as low per capita availability of water
due to increasing populations dependent upon fixed levels of
supply.

billion people will have to be met in order to provide them with safe
drinking water and to support water-borne sanitation systems. High water stress already affects 460 million people (8 percent of world population), and current water shortages and pollution cause widespread public health problems. Parallel negative effects on agriculture and economic
development, as well as on a variety of ecosystems, will severely impact
food supplies.

Water stress is due in large part to the expansion of irrigated agriculture, which contributes nearly 40 percent of world food production from
17 percent of all cultivated land. Thus, despite irrigation's profligate use
of water, it has often made it possible to alleviate food shortages brought
about by rapid increases in population.

The growing concern over the availability of global water resources
has been the impetus for a series of international conferences, such as the
International Conference on Water and the Environment, held in Dublin
early in 1992. This meeting adopted guiding principles for the management and planning of water resources at the local, national, and international levels (for the Dublin Water Principles, see Kay and Mitchell in
this volume).

The U.N. Conference on Environment and Development, held in mid-
1992 in Rio de Janeiro, concurred with the Dublin Conference. It called
for holistic management of freshwater supplies and the integration of
sectoral water plans and programs within the framework of national

economies and social policies. This encompassing approach to water resources management recognizes that water is a vital part of the global ecosystem and global socioeconomic activities.

The Middle East Setting

The protracted state of war between the Arab nations and Israel is fading and peace is gradually dawning. The Camp David peace accord between Egypt and Israel broke the cycle of war between the two states. More than a decade later another Arab-Israeli peace process was initiated. This latest process was started in Madrid in 1991 and resulted in a framework for a peace treaty between Israel and the Palestinians (the Oslo Accord, also known as the Declaration of Principles 1993). This was followed in 1994 by a peace treaty and normalization of relations between Israel and Jordan (see Appendix I for text of the treaty). The emerging atmosphere of cooperation, although slowed since 1996, will make it easier to tackle transnational environmental problems, the most prominent of which is water scarcity in the Jordan River basin.

Every state in the Jordan River basin is well below the global average of water availability (Engelman and LeRoy 1993; Commission on Sustainable Development 1997). As indicated by several chapters in this book, Jordan and Palestine are the most stressed riparians on the Jordan River. It has been argued that water stress of this nature will inevitably lead to "water wars." However, water shortages within Jordan, Israel, and Palestine, along with the transboundary nature of some of the sources involved (e.g., the West Bank Mountain Aquifer and the Jordan River), have led to proposed cooperative projects, designed to alleviate the resulting stress, which may emerge as testimonies to the benefits of cooperation and its contribution to greater water security, serving to move the peace process steadily forward.

It is natural to anticipate and experience triumphs and defeats on the road to Middle East peace. As we write this, the initial enthusiasm about the environmental and economic dividends of peace is waning. This is partly due to changes in the political landscape and the varying interpretations of such changes by the various actors. These recent developments contribute to the erosion of trust between Arabs and Israelis and make any cooperation, including that involving water sharing, extremely difficult. Currently, negotiations between Israel and Palestine have slowed, and those among Israel, Syria, and Lebanon remain frozen. Yet major documents of agreement have been signed and political precedents have been set. It can be argued that the tangible rewards of peace, the new global realities, and the repeated commitments of all parties to nonviolent resolution of the conflict make the peace process irreversible.

WATER IN THE MIDDLE EAST AT PEACE:
GEOGRAPHICAL PERSPECTIVES

"Why another book on Middle East water?" The editors are well aware of the wide variety of works on water resources of the Middle East in general and particularly on issues relating to the waters of the Jordan River. Nevertheless, we feel there are contributions yet to be made in three areas: (1) water resource studies from a geographical point of view, (2) forward-looking analyses of water resources management in light of the ongoing Middle East peace process, and (3) examination of the question, what will the changing management of scarce water resources in the Middle East look like as peace among the longtime antagonists is slowly established? These three areas define the approach, structure, and conceptual framework of this work.

The Geographical Framework

Geographical inquiry is a way of organizing knowledge about the human and physical environment in a systematic and ordered manner. Geographical investigation strives to provide accurate, orderly, and rational description and interpretation of the variable character of the Earth's surface (Hartshorne 1959). A variety of geographic frameworks are used in this book depending on the specific question at hand in each chapter. However, all are united by a geographic point of view which incorporates "inter-professional, inter-sectoral, and interdisciplinary dialogues needed to integrate land and water management" (Falkenmark and Lindh 1993, 88). Humans live and are part of the landscape, a landscape which is sustained by precipitation. People in turn access the natural resources by draining wetlands, irrigating fields, and building dams, canals, and pipelines, to name but a few of their activities. Interactions between natural elements and human actions may result in negative or positive impacts on environmental resources. Consequently, the sustainable management of the landscape is

> equivalent to a careful management of our interactions with it so as to meet three basic criteria for ecological security:
> 1. Groundwater must remain drinkable, land productive, and fish edible,

2. Biological diversity should be conserved,
3. Long-term overdraft of renewable resources should be avoided.
   (Falkenmark and Lindh 1993, 89)

This framework may be expanded by superimposing a political landscape on the environmental one. This superimposition demonstrates the
effects of human jurisdiction and sovereignties on the environment. In
the case of this book's subject matter, it emphasizes upstream/down-
stream geostrategic relations between riparians.

An upstream state may use its geostrategic position as leverage to advance its national or regional policy objectives. For example, it may unilaterally use a disproportionate share of water from an international
river. This may be done with relative impunity if it is militarily and/or
economically stronger than its downstream neighbors. Such an upstream
state may also threaten to limit the flow of water in order to advance
its foreign policy objectives. This deliberate manipulation of water flow
in response to the political climate between the involved riparians is
known as the "water weapon." Similarly, the export of water, particularly through permanent infrastructures, to a disadvantaged downstream state may create a situation where that state becomes dependent
upon and hegemonically influenced by the upstream one.

In other words, power politics usually favors advantaged upstream
states which may implicitly or explicitly dictate the terms of water management in the basin. Similarly, a disadvantaged water exporting state
may be intimidated by an advantaged importing state. Therefore, water
importers and exporters who are unable to maintain control over their
water resources (e.g., the rate and duration of flow) may lose the necessary flexibility for sustainable water management at the national level,
and as well find that their sovereignty is diminished. We refer to this
more complex construct as a "hydrostrategic landscape model."

MANAGING WATER SCARCITY: CHALLENGES AND OPPORTUNITIES

Countries in arid regions have many common challenges (see inset
"Challenges of Water Management in Arid Regions"). The distribution
of water resources varies spatially and seasonally within and between
states. A state may have abundant or sufficient water resources for its
needs at the aggregate national level, and at the same time have localities
that are water deficient. This spatial disparity between population con-

centrations and water availability necessitates expenditures in infrastructure, energy, and maintenance in order to reliably deliver water to population centers. For example, the Israeli National Water Carrier channels fresh water from the Sea of Galilee to water-deficient coastal cities.

Countries that are barely subsistent in water resources are vulnerable to the vagaries of drought and climatic change. Another common feature in water resources management is that the agricultural sector, on a world scale, consumes 87 percent of the fresh water withdrawn from the environment (Postel 1992).

Geography of Middle East Water: The Potential for Peace

It must be recognized that the location of and spatial variation in the physical environments of the Middle East affect the availability, consumption, and management of water resources. The location of political boundaries also results in the fragmented management of river systems.

The physical geography of the Jordan River riparian states is varied. Lebanon's northern mountains have precipitation levels that exceed 1,000 millimeters (mm) per year and receive large amounts of snow during the winter. On the other hand, over 50 percent of the land areas of Jordan, Israel, and Syria consist of hot, dry lands with precipitation levels below 250 mm per year. The eastern coast of the Mediterranean Sea is backed by a mountain range which declines in elevation from north to south. This creates an orographic effect making the coastline and the west-facing foothills relatively well watered and the interior dry. This spatial distribution of precipitation is most pronounced in Lebanon and Syria. In Israel and the West Bank, where the topography is reduced to rolling hills, the hydrological effects of orographic lifting are less evident. Overall, precipitation levels are highest along the western slopes of the mountain ranges that are parallel to the coastline, and decline in a north-to-south direction.

The human geography of the Jordan River basin, like that of the entire Middle East, is one of diversity. The overwhelming majority of the Israeli population is Jewish, that of Jordan and Syria, Muslim Arabs. While Muslims constitute a majority in Lebanon, Christians remain a significant minority.

On the other hand, there are important physical and human elements which serve to unify the core of this area. Lebanon and Israel are similar in that they both traditionally have had a Western orientation, relatively high incomes, very high levels of literacy, and low natural population growth rates. The Jordan Rift Valley (Ghor) is a physical feature that unites in a climatic, environmental, and physiographic way the people of Palestine, Jordan, and Israel. The latter two countries and Lebanon are similar in that they have substantial Palestinian populations, many occupying refugee camps. (Over 50 percent of Jordan's population is estimated to be of Palestinian Arab origin; about 20 percent of Israel's population, within its pre-1967 borders, is Arab.) Even linguistically there is similarity, with the population of the basin speaking variants of Semitic languages: thus, the resemblance between many of the words their languages share. "Peace," for example, is *shalom* in Hebrew and *salaam* in Arabic. Names of some geographical features are shown in Table 1.1.

Furthermore, as alternative and more felicitous political spatial patterns emerge with the unfolding of the peace process, many traditional and intervening obstacles will be erased (e.g., borders impermeable to trade and people, which have resulted in an absence of Israeli-Arab commerce). This will make it easier to deal with the approaching challenges of population growth, settlement of Palestinian refugees, and increasing water stress. Opportunities that can be realized only in a Middle East at peace will also be revealed. An example of this might be a large cooperative desalination plant capitalizing on economies of scale to reduce the unit cost of water, thereby alleviating water stress in Jordan, Palestine, and Israel. Similarly, any long-distance movement of water from countries such as Turkey will likely involve "have-not" Jordan River basin states. The more countries that benefit from such movements, the more the per-unit (or per-state) costs will be reduced. Greater participation may also encourage the financial assistance from, among others, oil-wealthy Arab nations. Therefore, broader regional spatial and financial linkages could be realized, thereby positively affecting the management and redistribution of water in the Middle East.

The similarity of physical and human features mentioned above will make it easier to understand and manage fluctuations in climate and resource flows within the region's ecosystem. Moreover, members of the Palestinian Arab minority in the Jewish state of Israel are uniquely positioned to play a key role in mediation between the Arab states and Israel, or as interpreters of Arab culture and tradition to Israelis, as well as Jewish culture to the Arab world at large. This could well break down existing barriers and expedite cultural and economic interactions. Cooperative projects in the areas of research and development—some of which are already under way—are among the first joint ventures between potential peace partners. This will help with the gradual diversification of

CHALLENGES OF WATER MANAGEMENT IN ARID REGIONS

Water Quantity

THE AVAILABILITY OF water resources can be classified into the following spatial scales. At a macro (regional) scale, water is generally scarce across the Middle East. This is particularly true in the region south of the Fertile Crescent (which is delimited by the Euphrates-Tigris, Asi [Orontes], Litani, and Jordan Rivers, and by the 250-mm isohyet south of which rain-fed agriculture cannot exist). At a meso (national) scale, some countries such as Turkey and Lebanon have more water today than they are likely to consume in the near future. Other states such as Jordan and Israel presently are consuming about 15 percent more water than is annually renewable. At a micro (provincial or farm) scale, the situation is somewhat paradoxical. Too much water is being consumed in certain sectors or provinces relative to its meager availability at a national scale.

In addition to the national challenge of scarce water resources, water in the Jordan River basin states must often be moved, sometimes over long distances and across difficult terrain, from areas of high precipitation and low population to areas of low precipitation and higher population concentrations. Jordan's East Ghor Canal and Israel's National Water Carrier are two prominent examples of this. Such transfer of water is made possible by high investment in energy and capital for start-up and maintenance.

Water Quality

ALTHOUGH TENSIONS over water in the Middle East have historically been over quantity and allocation, concerns over water quality are increasing. There are two principal sources of water pollution: point sources, where pollutants are discharged from easily identifiable industries and from sewage plants which remove some but not all pollutants, and non-point sources, which are difficult to trace. For example, Palestinian communities on the West Bank complain that their water supply is being polluted by wastewater from specific urban Israeli settlements located on higher ground. Non-point sources of pollution originate from spatially diffuse areas where water is polluted by runoff, subsurface flow, or deposition from the atmosphere. Israel is concerned about how Palestinian economic activities—including agriculture—on the West Bank may affect the replenishment and/or quality of the westward-flowing Mountain Aquifer. Similarly, Syria alleges that its reduced water flow in the Euphrates is being polluted by agricultural return flows from southeastern Turkey. The former issue is likely to be addressed in the final peace accord between Israel and Palestine.

Not all concerns about pollution fit neatly into these categories. Israel's Coastal Aquifer, overexploited since the early 1960s, experiences varying levels of salination. Consequently, it has been artificially recharged primarily during the winter months since 1964, although those portions totally invaded by seawater cannot be reconstituted.

Variability

THE MANAGEMENT of water resources must adjust to the high variability of the intricate and dynamic hydrologic cycle, which is affected by human activity and (mis)guided by varying ideologies and government policies. This results in externalities due to the interdependence of the natural and human systems, and the (over)utilization of surface and subsurface sources. Consequently, it is difficult to manage water resources using only pricing and market forces.

Fragmented Planning and Management

VARIOUS TYPES OF water uses and issues related to the quantity and quality of water are typically managed by separate government departments or agencies. In other words, irrigation, hydroelectric power, and municipal water supply are often managed by independent ministries or agencies which do not coordinate their activities.

International rivers are usually seen as sources of water or sinks of pollution, and not as unified interconnected ecosystems (see Kolars in this volume). Segmented management of such water bodies may lead to inefficiencies, ecosystem degradation, environmental injustice, and sometimes conflict. Such an approach to resource management undermines sustainable utilization of international rivers.

Water Pricing

IRONICALLY, FARMERS in some arid countries such as Saudi Arabia, Syria, and Israel receive subsidized water. This subsidization to farmers as well as to urban residents was cited by the Israeli State Comptroller as a major reason for the current water crisis in that country (Kliot 1994). The provision of water at prices far below its economic value is often due to well-organized and politically influential farm lobbies, and in other instances can be traced to ideological convictions. Thus it is often politically expedient to maintain or increase supplies. Water underpricing, therefore, serves to inflate supply rather than reduce demand. This bias is reflected in water-intensive cropping in arid regions such as the American Southwest and the Jordan River region.

Globally, governments are estimated to subsidize the use of energy, water for agriculture, and passenger road transport to the tune of US$700 billion (*Building the Transition* 1997). A recent World Bank study reviewed the price charged for municipal water supply in projects that it had financed. It found that the price charged covered only about 35 percent of the average cost of supply. Farmers in the Middle East and other arid areas are widely believed to pay an even lower percentage of the cost of water.

Costly New Sources

COUNTRIES TYPICALLY tap their lowest-cost and most reliable sources of water first. As these sources are fully utilized, the development of new sources carries with it heavier financial and environmental costs. The consequences of water stress stemming from the dominant use of water for agriculture are likely to be greatest for economies where the agricultural sector is a major employer and a significant contributor to the gross domestic product.

For example, Jordan's population increased from 0.6 million in the early 1950s to 3 million in the 1980s, and is expected to reach 7.4 million in the year 2015. In addition to this, rapid urbanization in the kingdom and rising incomes will boost municipal and industrial water consumption from about 25 percent of the total water used in the early 1990s to about 40 percent in 2015. Population pressures have forced the government to irrigate an additional 48,500 hectares and to invest in technologies that are introducing greater efficiency to the agricultural system. Jordan can increase its water supply by (1) building a reservoir jointly with Syria on the Yarmuk, (2) tapping the Disi Aquifer in southeastern Jordan, and (3) treating a much greater volume of wastewater.

In addition to the heavy financial burden attached to all these options, other complications include environmental and political challenges (e.g., mining the nonrenewable waters of the Disi Aquifer, which is shared by Saudi Arabia). Jordan may also use the waters of the Yarmuk, which it shares with Syria and Israel.

Noncooperative Regimes

DATA ARE NOT shared freely among riparian states; this is true among riparians along the Jordan River. There is often considerable confusion among riparians over the meaning of such concepts as water availability (potential, dependable, readily available, regionally vs. locally available), water accessibility (what could be available locally or what is actually made use of by the local population), and land area units (e.g., the area of a dunam, a feddan). Riparians, moreover, often use different measurements and methodologies in collecting data.

TABLE 1.1. Some Common Arabic and Hebrew
Geographical Terms

Arabic	Hebrew	Meaning of Term
'ain	'ain	water spring or fountain
ari'd	eretz	land
ma'	mayim	water
nah'r	nahar	perennial stream
wadi	vadi	ephemeral stream
beit	beit	home
raas	rosh	geographic head, cape
lisan	lashon	geographic tongue, peninsula
tell	tel	hill
yamm	yam	sea
khirbat	khirbat	archaeological ruin
ghamam	'anan	cloud
shamis	shemesh	sun

the economy away from water-intensive crops. By the same token, low population birth rates in Lebanon and Israel may provide models for other riparians.

Adding to the potential for regional cooperation, countries in the Jordan River basin are gradually supplanting wasteful supply-based

OPPORTUNITIES FOR WATER MANAGEMENT
IN ARID REGIONS

Demand Management

DEMAND MANAGEMENT is achieved through tariff regimes, effective methods of revenue collections, consciousness raising regarding water scarcity, and encouraging water saving strategies in all sectors of the economy. All these boost water availability in time and space. The Food and Agricultural Organization of the United Nations recommends that water planners need to consider balancing competing demands spatially (rural/urban), hydrologically (upstream/downstream), and politically (allocation of water between provinces or riparians). Planners need to also preserve the ecological integrity of the basin's ecosystem.

According to recently enacted World Bank principles, riparians should give priority to:

1. household water security,
2. food security,
3. water quality and human health,
4. environmental stewardship, and
5. regional cooperation.

Use Subsidization

GIVEN THAT MANY low-value crops are often irrigated, the existing subsidy-based system encourages misallocation of a scarce resource. Beaumont (1994) predicts that as hydrological stress intensifies, low-value agricultural uses per cubic meter of water will be replaced with higher-value domestic and industrial uses. The Economist (28 March 1997, p. 11) argues that water "needs to be diverted from farms to towns." The United Nations' Commission on Sustainable Development (1997) recommends that countries with high water stress should redirect their water resources toward the production of high-value products that are not water-intensive. It also argues for the need to add value to national agricultural products by developing related processing industries, and to generate greater foreign exchange that might be used for food imports.

Appropriate Regulations

REGULATIONS ARE needed to minimize the impacts of pollution, especially from non-point sources, on national and international watercourses.

management practices with conservationist demand-based use of water (see inset "Opportunities for Water Management in Arid Regions"). At the same time, in a Middle East at peace, it will be easier to identify and minimize the effects of point and area sources of pollution.

THE FORMAT OF THIS BOOK

Our chapters recognize the inevitable path toward peace and are organized chronologically from the past to the future. Lonergan and Beaumont each provide an overview of the Jordan basin and its riparians, while Wolf gives the historical setting of water and conflict on the Jordan River, as does Amery for the Litani. We then look toward the immediate future—Hof describes what the water issues may be when Syria and Israel begin negotiating, and Kay and Mitchell describe the "hydrosecurity" needs of each riparian and the difficulty of planning for uncertainty. Kliot provides details of a cooperative framework for sharing water resources, while Rowley describes the political constraints on such a framework. Finally, Kolars reminds us of the needs of one party so rarely represented at the negotiating table, the river itself.

While each chapter describes various aspects of hydrology, history, and political needs, the questions and approaches to answers are essentially geographical: What is the relationship between people and their natural resources? How are geopolitical needs manifested in the landscape? Can generalizations be made about watershed management when each basin's hydrology, politics, and culture make, as Gilbert White pointed out more than forty years ago, each river unique?

Each chapter adds its author's (or authors') perspective in answering these questions. Peter Beaumont and Stephen Lonergan provide an overview of the Jordan River watershed and the historic and future needs of its riparians. The former's emphasis is on the needs of Israel, Jordan, and the West Bank Palestinians. Beaumont argues that the greatest threat to cooperative water management will be population growth and the subsequent increase in demand. He describes Israel's neglect of water infrastructure during decades of occupation of the West Bank and Gaza. The economic value of water in Israel is then described, as well as technological advances, particularly in wastewater reclamation. He leaves the reader with a question for the future: conflict or cooperation?

Lonergan analyzes the links between water scarcity and security. He points out that three interconnected crises indicate the complexity of water security issues: water supply and demand, deteriorating water quality, and geopolitics. He then applies Gleick's typology of resource-related conflicts to Middle East hydropolitics, linking issues of global change—population growth, urbanization, and climate change—to growing difficulties in water management. Like Beaumont, Lonergan looks to water marketing as a possible solution, but warns that market forces cannot be implemented until water rights are clearly allocated.

Wolf turns to historical geography, seeking to mitigate popular charges that much of the Arab-Israeli conflict was fueled by "water wars." He assesses historic boundary development, warfare, and negotiations and concludes that, while water has influenced the shape of boundaries, particularly between the British and French Mandates, no territory was ever explicitly targeted, captured, or retained because of its access to water resources. Furthermore, in ongoing negotiations, water has not been used as an excuse for retaining territory. Joint management, rather, has been favored over territorial sovereignty.

Amery examines the claims of Israeli-initiated water diversions from the Litani River of Lebanon. He argues that historical factors, militarily and politically reconfigured space, and the creation of off-limit areas have all contributed to the development and permanence of the popular hypothesis that the Litani River has been diverted, in part, into the headwaters of the Jordan. Decades of Israeli occupation and of tensions in southern Lebanon make it very difficult to independently gather conclusive evidence about the possible existence or nonexistence of such a diversion. The sometimes contradictory and vague evidence provided by supporters and detractors of the hypothesis, Amery argues, makes it difficult for anyone to make conclusive assertions about the issue. Given this uncertainty, he believes that the weight of the evidence rejects the existence of large diversions, but leaves open the possibility of small-scale transfers. Amery's conclusion is that irrespective of whether water diversion actually exists, it certainly exists in the minds and discourse of local inhabitants, and must therefore be addressed. Consequently he suggests that when the Middle East is finally at peace, public participation in environmental planning can help build confidence between the two peoples.

Fred Hof focuses on the immediate future and describes what the water dimensions will be when Israel and Syria negotiate a settlement of the sovereignty of the Golan Heights. He suggests that four geographical issues will be particularly sensitive—the Banias Springs, Golan surface water, the upper Jordan River, and the Sea of Galilee. After a summary of how past negotiations have influenced current positions, Hof outlines his view of what a settlement of the Golan negotiations might look like for each of the four water-related issues, and suggests that each side might be able to achieve its hydrostrategic goals in such an agreement.

Kay and Mitchell also look to the future and broaden the discussion by linking the interests of each riparian, as expressed through the concept of water security, with performance criteria and attention to existing uncertainties. They lay out in their conceptual framework performance criteria used to evaluate sustainability, including the reliability, resiliency, and vulnerability of the system. They then apply these criteria to a performance evaluation of Israel's water system and describe water's role in national management policies, and in the peace instruments which have been signed. They conclude that traditional management of water resources in the Jordan basin is precarious and unsustainable and suggest alternatives through adaptive management criteria.

Nurit Kliot takes a more institutional approach to the geopolitical interests of the Jordan riparians, offering a cooperative framework for sharing water among Israel, Jordan, and the Palestinian Authority. She presents a brief overview of the institutional history of the riparians, including a description of each of the peace documents signed to date. She then describes institutional frameworks which have been used to manage shared water resources, citing legal doctrines, the recent experience of the International Law Commission, and other basin accords. She suggests that the Jordan River Accords go beyond simple coordination to include the implementation of some cooperative projects, but fall short of true integrated watershed management. Kliot concludes that, as in other basins, water wars can be avoided, provided the political will exists to continuously seek creative solutions.

Of the contributors, Gwyn Rowley is the most critical of the peace process, and his chapter offers a note of skepticism regarding the road toward peace. He wonders whether the necessary and difficult compromises are actually feasible, given historical intransigence. He begins in the past, describing water as a strategic resource, particularly as it pertains to the historic tensions between Arabs and Israelis. He then describes ongoing attempts at the creation of international water law, and suggests that such law, still in its infancy, is not sophisticated enough to offer the specific guidelines necessary for resolution of the Jordan River dispute. This lack of legal guidelines, he argues, is compounded by Israeli territorial and hydrological aspirations in the region which threaten the possible benefits of cooperation between riparians.

Finally, John Kolars speaks for the one party to the dispute without a voice—the river itself. Kolars reminds us that, even as negotiators strive to bargain away the entire flow of a river, instream ecology has its own demands and needs. He argues, moreover, that the river holds critical lessons for us, that the relationship between people and the water they use is truly symbiotic. Kolars translates a need for river advocacy into the realpolitik world of Middle East hydropolitics and offers applications for both the Jordan and the Tigris-Euphrates systems.

Our focus on the Jordan necessarily precludes detailed discussion of many topics of import in the region, such as the intertwined hydropolitical, legal, cultural, and historical aspects of other nearby watersheds—the Tigris-Euphrates, the Nile, and the Orontes. Likewise, the most critical scarcity of water in the region, that in the Gaza Strip, along with the attendant health hazards, is not included.

A note on usage: Language, like most issues in the Middle East, is highly charged. The term one uses for a simple place name can connote intricate and varying political viewpoints. West Bank, Occupied Territories, and Administered Areas all refer to the same geographic entity, as do Lake Tiberias, Sea of Galilee, and Lake Kinneret. Given such a setting, we, as editors, simply did not feel comfortable mandating a unified format to the authors. The reader may note, then, some inconsistencies from chapter to chapter in usage.

Finally, we note that one of the first documented water disputes in history—that between Isaac and the herdsmen of Gerar (Genesis 26: 19–22)—was resolved when Isaac dug his well elsewhere. The Middle East is out of "elsewheres." While population growth and rising standards of living are driving water demand inexorably upward, the natural supply is roughly what it was in Isaac's day. While Malthusian doomsday scenarios continue to be pushed back with advances in technology and social organization, the fact remains that unilateral development of international rivers has reached its limits. The only rational solutions are cooperative ones; the only boundaries which belong on the planning map at this stage are those of the watersheds themselves.

When asked, "What will the water issues of the Middle East at peace look like?" we in this book answer from the perspective of geography. We believe that the spatial and integrative nature of geography sheds a much-needed light on the social, economic, geophysical, political, and environmental variables that affect the management of limited water resources.

We also seek full recognition of the difficult history which led to this point, and the dangers which lie ahead. The chapters in this work do not

ignore the difficulties in achieving truly integrated watershed management—they agree only on its inevitability.

REFERENCES

Beaumont, P. 1994. "The Myth of Water Wars and the Future of Irrigated
Agriculture in the Middle East." *International Journal of Water Resources Development* 10 (1): 9–22.

*Building the Transition to Sustainable Development: A Critical Role for the
OECD.* 1997. The Report of the High Level Advisory Group on the Environment, 25 November. http://www.oecd.org/subject/sustdev/index.htm.

Clarke, R. 1991. *Water: The International Crisis.* London: Earthscan.

Commission on Sustainable Development. 1997. *Comprehensive Assessment
of the Freshwater Resources of the World.* New York: United Nations.

Engelman, R., and P. LeRoy. 1993. *Sustaining Water: Population and the Future of Renewable Water Supplies.* Washington, D.C.: Population Action
International.

Falkenmark, M., and G. Lindh. 1993. "Water and Economic Development."
In P. H. Gleick (ed.), *Water in Crisis: A Guide to the World's Fresh Water Resources,* pp. 80–91. Oxford: Oxford University Press.

Falkenmark, M., and C. Widstrand. 1992. "Population and Water Resources: A Delicate Balance." *Population Bulletin* 47 (3): 1–36.

Hartshorne, Richard. 1959. *Perspectives on the Nature of Geography.* Chicago and London: Rand McNally.

Kliot, N. 1994. *Water Resources and Conflict in the Middle East.* London
and New York: Routledge.

Postel, S. 1992. *Last Oasis: Facing Water Scarcity.* New York: W. W. Norton
and Co.

United Nations. 1992. *Water Resources of the Occupied Territories.* New
York: United Nations.

2. *Conflict, Coexistence, and Cooperation*
A STUDY OF WATER USE IN THE JORDAN BASIN

Peter Beaumont

INTRODUCTION

In a world context the Jordan River is a small basin with an area of only 18,300 square kilometers. However, because of its associations with three of the world's great religions, Christianity, Islam, and Judaism, it has become one of the best-known rivers in the world. At the present day four countries, Lebanon, Syria, Jordan, and Israel, share its waters (see Figure 2.1). Another group of people, the Palestinians, also have just aspirations for a state of their own within the Jordan basin. This chapter examines the pressures which now exist on the water resources of the Jordan River and the ways in which the resources might be utilized in a more efficient manner if mutual trust and understanding amongst the countries can be developed as a result of the peace process begun in Oslo in 1993. Inevitably most attention will be focused on Jordan, the West Bank, and Israel, as these are the vital entities in terms of their control of the waters of the Jordan basin.

HISTORICAL DEVELOPMENT

The countries which exist in the region have boundaries which were created as a result of the breakup of the Ottoman Empire following the end of the First World War. The boundaries which came into existence were largely artificial and reflected the relative strengths of the great powers at that time. In the north, Lebanon and Syria developed under the French Mandate, while to the south, under the British Mandate, Palestine and Transjordan came into existence. At the end of the First World War the four mandated territories possessed weak economies based largely on subsistence agriculture. Urban centers, such as Jerusalem, Damascus, and Aleppo, provided marketplaces and a range of craft industries, but trade on anything but a regional basis was minimal.

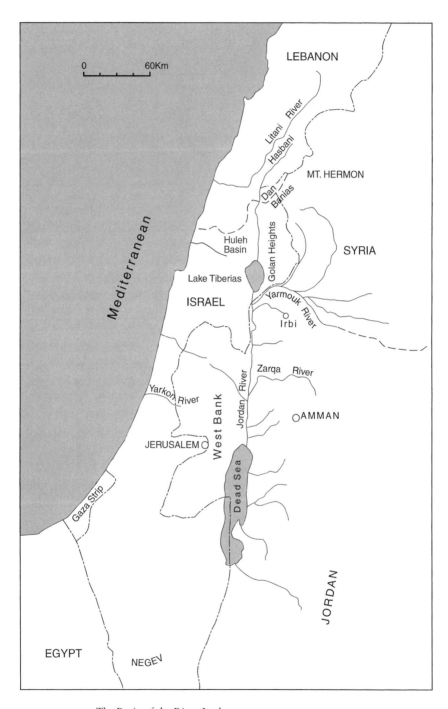

FIGURE 2.1. The Basin of the River Jordan.

Population numbers were low, but are difficult to quantify. In the 1920s the population of Palestine was probably around 750,000 (Beaumont, Blake, and Wagstaff 1986, 433). In Transjordan no population estimates are available before 1940, when it was estimated that the population was between 300,000 and 350,000 people. In Lebanon estimates suggest a population of around 840,000 in 1920, while for Syria the total population then was probably around 2 million people. In all, the population of these four future nations was less than 4 million. Standards of living were almost everywhere low, except for a few urban centers.

Water usage rates were also low at this time. In urban regions water supply would be from wells, or if available, from surface sources. In the northern part of the region agriculture was largely rain-fed, but to the south irrigation became more important as precipitation decreased. Although serious water shortages could occur locally in drought years, in general the lack of water was not a limitation on economic activity. Indeed, centers such as Damascus were famous for their abundance of water supplies.

In the post–Second World War period one of the key features of the region has been the very rapid growth in population numbers. By the mid-1990s Lebanon had a population of 3.7 million, Syria 14.7 million, Jordan 4.1 million, the West Bank 1.5 million, and Israel 5.5 million. This total of 29.5 million people is approximately seven times the figure for the same countries in the early 1920s and approximately double the figure for the early 1970s. This rapid population growth has put tremendous pressure on both water and food resources. The cultivated area devoted to cereals has increased markedly, but nearly everywhere yields have remained low. The only exception to this has been in Israel. By the beginning of the 1960s almost all the easily cultivated land was being utilized. At the end of the 1960s these four countries, and many others in the Middle East, which had been self-sufficient in basic foods such as cereals, now had to import wheat in ever larger quantities (Beaumont 1989). The era of food self-sufficiency was over forever.

The growing pressure on water resources mirrored that on food, as a large proportion of the water was used for irrigation. Almost all the water utilized came from surface water sources, as there had been little tradition of groundwater use except for domestic supply. Once again, though, it was not until the 1960s that the prospect of serious water deficiencies began to become apparent. Israel was the first country to experience such difficulties. Other countries in the region are now beginning to attain full usage of their easily available water resources.

Since 1948, when the state of Israel was established, the Jordan basin has been the scene of numerous conflicts. A state of war existed between Israel and its Arab neighbors, and this period was punctuated by active hostilities in 1967, 1973, and the early 1980s. By the late 1970s, with the signing of the peace agreement between Israel and Egypt, the period of open conflict ended and was replaced by a watchful and guarded coexistence. The formal acknowledgment of a change from a state of conflict to one of coexistence has been slow to take place. Jordan only signed a peace treaty with Israel in 1994 and Syria has yet to do so. However, few on either side of the political divide believe that full-scale open conflict will occur again. Instead it is hoped that the uneasy coexistence which currently prevails between Israel and its neighbors will flower into a coexistence based more on mutual trust.

Cooperation amongst the different entities still seems a long way off, particularly as the Arab nations often disagree amongst themselves. Initially, cooperation will be between the security forces of the Israelis and the Palestinians, but it is still likely to be a long time before true economic cooperation takes place. Water is a resource which would benefit from a unified policy of development, but agreement on this seems a long way off. Since 1948 the patterns of water use in and around the Jordan basin have seen developments taking place solely on a national basis. However, ideas of unified development of the water resources of the region, including the Litani and the Jordan, date back to the 1930s or even earlier (Wolf 1995). Unfortunately, most of these approaches were based on a Zionist perspective, and very little, if any, attention was paid to the needs of the Arabs. Perhaps the best-known work of this kind is the book *Palestine: Land of Promise* by the American Walter Clay Lowdermilk, which was published in 1944.

JORDAN: WATER NEEDS AND AVAILABILITY

The country of Jordan remains one of the poorest countries of the Middle East, with a per capita GNP in 1993 of only US$1,190 per annum. Even so, it has experienced considerable changes since the end of the Second World War. One of the biggest of these changes has been the growing urbanization of the population as traditional agriculture has been unable to support the extra people in rural environments. By the mid-1990s the urban population had reached a figure of 68 percent of the total of 4.1 million (Population Reference Bureau 1995). Associated with this urbanization has been the growth of the service sector, with the result that slightly more than 60 percent of the total work force are now classed as service-oriented. Industrial employment is around 27 percent and agricultural about 11 percent. This latter figure is somewhat misleading, as there remains a large and informal input of labor into the agricultural sector which does not appear in official statistics. As yet the industrial sector is only poorly developed, with the major emphasis on large-scale fertilizer production. However, small-scale industrial production is growing.

Overall, Jordan is a very dry country. The highest precipitation, in excess of 600 mm, occurs in a narrow belt along the highlands overlooking the rift valley. Almost all of this precipitation falls during the winter season. In the Jordan valley annual precipitation totals are usually less than 250 mm, while over most of the southwest of the country annual totals are less than 50 mm. Overall, 80.6 percent of the country receives less than 100 mm. In an average year the total volume of precipitation falling on Jordan is about 7,200 million cubic meters (Salameh and Bannayan 1993).

The surface water resources of Jordan are located in the two major rivers, the Jordan and the Yarmuk, and in a large number of smaller streams draining the well-watered western highlands. Excluding the waters of the Jordan and the Yarmuk, these streams, including the Zerqa, have a total discharge of around 349 million cubic meters each year. The fact that there are so many small streams along the highlands overlooking the rift valley meant that water availability was not a serious problem for most settlements before the 1980s.

The Yarmuk, which is the largest tributary of the River Jordan, is shared mainly between Syria and Jordan, with most of the discharge being generated in Syria. In recent decades its flow appears to have declined significantly. Between 1927 and 1954, it is claimed, the discharge averaged 467 million cubic meters per year. This had fallen to a figure of 400 million cubic meters per year in the period 1950 to 1976, and more recent data suggest a further fall to around 360 million cubic meters per annum (Salameh and Bannayan 1993). However, with the many water resource projects which have been developed in recent years within the basin, it is difficult to be certain what is the result of natural precipitation changes and what the impact of human development. At the present time Syria is believed to be abstracting 160–170 million cubic meters each year, Jordan 100–110 million cubic meters, and Israel around 100 million cubic meters (Salameh and Bannayan 1993).

The Zerqa catchment covers an area of 4,025 square kilometers and is by far the most urbanized river system in Jordan. Indeed, the catchment houses about two-thirds of the population of the country and approximately four-fifths of the industry. As a result pollution potential is high. The natural flow of the river was around 65 million cubic meters each year (Salameh and Bannayan 1993, 21). However, since 1977 the river discharge has been altered by the construction of the King Talal Dam.

There is also a major import of water into the Zerqa basin to supply the Amman-Zerqa conurbation. This amounts to around 40 million cubic meters per annum and is derived from Azraq and groundwater sources in the south. With increased water use in the basin the volume of effluents with high levels of dissolved solids has substantially increased also. This was particularly apparent in 1985 with the opening of the Khirbet es Samra wastewater treatment plant. Such has been the impact of its effluent on the River Zerqa that, since its construction, during some periods the water which is stored in the King Talal reservoir is not suitable even for irrigation. Concern is also being expressed that this low-quality water may well be beginning to pollute the aquifers over which the River Zerqa flows. Currently it is estimated that effluent flows make up about one-half of the flow of the river. This means that in the summer months the river water quality is extremely low, though in winter the quality improves as a result of dilution of the effluents with flood flows.

The groundwater resources of Jordan are found in three main aquifer complexes (Salameh and Bannayan 1993). The oldest is a sandstone complex with rocks of Paleozoic and Mesozoic ages. In the south of Jordan it forms a single unit, but in the north it is separated into two aquifer systems by limestones and marls. The older and deeper unit is known as the Disi Group Aquifer System. This underlies most of the country of Jordan, but only outcrops in the south along the rift valley region. The main flow direction of the water in this aquifer system is toward the northeast. The upper and younger unit is known as the Kurnub and Zerqa Group Aquifer System. This outcrops in the lower Zerqa basin, and along the escarpment of the rift valley. Directions of flow are more complex in this aquifer than in the Disi system. In the south the dominant flow direction is to the northeast, in the central part of the country it is to the west, and in northern Jordan it is to the southwest.

The Upper Cretaceous aquifer system has a maximum thickness of 700 meters. Two major formations make the main aquifers, although series of smaller ones also occur. The lower formation is made up of a crystalline limestone and is characterized by high porosity and permeability. It outcrops widely in the northern highland region, and this is the main source of water recharge. To the east the aquifer is covered by impermeable marls and limestones. In turn, these deposits are overlain by an upper formation which forms the best aquifer in the group. It outcrops in the highland areas and here recharge occurs. Eastward, the upper aquifer is buried by marls and like the lower formation becomes a confined aquifer. Groundwater flow in this aquifer system is complex. In general, to the west of the topographic divide, water flow is toward the west. However, elsewhere the majority of the flow is to the east.

The third group is classed as shallow aquifers, and these form two main systems. The first system is provided by basalt formations which extend from the Jebel Druze in Syria southward to Azraq and the Wadi Dhuleil. Recharge occurs on the slopes of the Jebel Druze, and the groundwater flows radially away from this highland massif. However, geological structures concentrate the water flows into the upper Yarmuk basin, the Wadi Zerqa basin, and the Azraq basin. The final group is made up of alluvial deposits and sedimentary formations of Tertiary and Quaternary ages. They are of widespread surface occurrence, but tend to be of very small scale. Recharge of these aquifers takes place directly from precipitation or sometimes from underlying aquifers. Groundwater flow directions are largely determined by local relief.

The groundwater resources of Jordan can be divided into those which are being replenished under present-day conditions and those which were put into storage during wetter conditions in the past. This latter water is essentially "fossil" in nature, and once the water is removed it will not be replaced. Such resources tend to be located in the drier eastern and southern parts of the country. The available groundwater resources are estimated to total 340 million cubic meters per annum.

The total renewable water resources of Jordan, including groundwater, are estimated at around 900 million cubic meters per annum (Salameh and Bannayan 1993, 172). However, it has been suggested that storm runoff in some years might produce up to a further 340 million cubic meters (Kliot 1994, 226).

The growth of water utilization has been rapid in recent years. In the early 1960s irrigation usage was probably around 250 million cubic meters per annum (Kliot 1994, 228). By the late 1970s this had grown to around 400 million cubic meters per annum, and by the early 1990s it was approaching values of 650 million cubic meters (Salameh and Banayan 1993). In the domestic sector the pattern of growth of water use was even more rapid. In the mid-1970s demand was around 40 million cubic meters per annum, but by the early 1990s it was 180 million cubic meters per annum (Kliot 1994; Salameh and Bannayan 1993). Until recently the industrial demand for water had been low, but by the early 1990s it rose to around 45 million cubic meters per annum. From the above figures it can be seen that total water consumption in the early 1990s was 875 million cubic meters per annum. This, it will be noticed, is approximately the same value as the estimated renewable water resource base of the country. By the year 2000 irrigation water demand in Jordan is expected to reach 720 million cubic meters, domestic demand 340 million cubic meters, and industrial demand 60 million cubic meters (Kliot 1994, 231). This adds up to 1,120 million cubic meters—a figure which is 200 million cubic meters in excess of what is thought to be the renewable water resource base of the country.

At the present time overpumping of aquifers is already occurring and is believed to be reaching values of 120 million cubic meters per annum in excess of recharge (Salameh and Bannayan 1993, 104). Most of this overpumping is associated with fossil water in the deeper aquifers, but it also takes place around larger population concentrations, such as in the Amman-Zerqa region. It is occurring because it is extremely difficult to utilize efficiently many of the surface water flows of the smaller wadis. As a result mining of groundwater is considered to be an easier way of obtaining needed water resources.

The biggest problem facing Jordan is the growth of its population and the impact that this will have on the demand for water. Equally important is the fact that at the moment domestic water demand in Jordan is low, averaging only about 44 cubic meters per capita per year. This low figure of domestic water consumption means that as standards of living rise growth in per capita water demand will also impose greater strains on the available water resource base. One of the present ironies with urban water use in Jordan is that the poorest people, who have to have their water delivered by tanker, as they are not connected to main supplies, have to pay around US$3 per cubic meter for their supplies. This is well in excess of what richer members of society pay for their piped water supplies.

It is not unreasonable to expect that urban water demand in Jordan will rise from the mid-1990s value of just over 44 cubic meters per capita per year. If it were to reach a figure close to 100 cubic meters per capita per year in the early decades of the twenty-first century, the implications in water resource terms could be considerable. This latter figure is similar to consumption in Israel, which is Jordan's neighbor. It would mean that urban demand could rise from a figure of 180 million cubic meters per year in the early 1990s to a value of 620 million cubic meters in 2010 and 830 million cubic meters in 2025. This figure of 830 million cubic meters per annum, based on a predicted population of 8.3 million, is getting very close to the 900 million cubic meters per annum which is considered to be the total renewable water resource base for the nation. If this estimated figure proved to be correct it would mean that the water left over for both industrial and agricultural use would only be around 70 million cubic meters each year.

However, these postulated high domestic water demands seem unlikely to occur, as the large population growth rates will ensure that a sizeable proportion of the country's population will continue to be very poor. In the richer urban areas per capita demand will undoubtedly reach and perhaps even surpass 100 cubic meters per person per annum, but elsewhere demand will be much lower. However, a conservative estimate of domestic demand around 55 cubic meters per capita per year will still produce a total urban water need of 545 million cubic meters for the predicted 2025 population.

The fact remains, though, that with a total renewable water resource base of around 900 million cubic meters per annum the only way in which domestic demand can be met is at the expense of agricultural water usage. If the low value of 545 million cubic meters per annum for domestic water demand in the year 2025 is approximately correct it means also that agriculture and industry will have to survive on about 355 million cubic meters per annum. In the early 1990s industrial use was relatively low at a figure of 45 million cubic meters each year. Growth here is difficult to predict, but with population increase and the need for new food processing industries it seems possible that demand might double by 2025 to around 100 million cubic meters per annum. In effect, this means that irrigation will be left with only 255 million cubic meters per annum, compared with a figure of 650 million cubic meters in the early 1990s. This seems likely to be a maximum estimate of agricultural water availability, as the urban water demand might increase more rapidly than the modest increase postulated above.

Up to the present time relatively little attention has been paid to the reuse of wastewaters following treatment. It has been claimed that up to 60 percent of urban water consumption in Jordan can be reclaimed and reused. However, estimated costs for collecting and treating wastewaters in Jordan are already quite high at US$0.37 per cubic meter (Haddadin 1996, 68).

The only other option which Jordan possesses is to increase the rate of groundwater mining. There is little doubt that in the short term this figure could be increased, but it is equally certain that water quality would decline and eventually saline water would be pumped. At the moment in parts of south Jordan high-quality water is being squandered on wheat production. Under irrigated cultivation, wheat yields can reach 4 tons per hectare. However, the hot summer temperatures require water application rates of around 10,000 cubic meters per hectare. This means that each ton of wheat needs 2,500 cubic meters of water (Beaumont 1994). The cost of wheat on the open world market varies, but is usually between US$150 and 200 per ton. What is interesting is that if the water used for irrigation had to be obtained from other sources its cost would be much higher. For example, if it were replaced by desalinated water, which costs around US$2 per cubic meter, the 2,500 cubic meters required to grow a ton of wheat would cost US$5,000. In other words, every time a ton of wheat is grown by irrigation it is a loss to the country of about US$4,800 in terms of the replacement value of the water for other uses.

The problem for the Jordanian government is that at the present time no other major demand for the groundwater which is being used for wheat production exists. Therefore, the farmers are content to go on using the water. As far as they are concerned the cost of the water is how much money needs to be spent to pump it out of the ground. The fact that the water is being mined and that once it is used up it will never be replaced is of little significance to them. Quite naturally all they are interested in is maximizing their profits. What needs to be done is for the Jordanian government to devise a system to stop fossil water being used for low-value purposes such as wheat production. If this could be achieved it would mean that the water would be available in the future for more valuable uses such as water supply for Amman, Aqaba, and other major cities.

ISRAEL: CHANGING WATER NEEDS

The pattern of water availability in Israel follows a pattern very similar to that of Jordan. In general, precipitation amounts decline toward the south and are highest in the highland regions. The biggest difference is that in Israel the highlands are located in the east of the country and to the west there is only a short distance between the highlands and the Mediterranean Sea. This coastal plain area is the most fertile part of Israel, and it is here where most of the agricultural development has taken place since the formation of the state. In the north cultivation is possible using dry-farming techniques, but farther south irrigation is needed to produce assured crop yields.

The highest precipitation totals in Israel range from 500 to 750 mm in the north, while in the south values fall to around 50 mm a year. The total precipitation falling over the country averages between 8,000 and 10,000 million cubic meters each year (Shahin 1989). Approximately two-thirds of this total is lost by evaporation and evapotranspiration.

The surface water resources of Israel are dominated by the waters of the Jordan River, which forms the western border of the country for part of its length. Almost all of the other rivers rise in the highlands and flow westward to the Mediterranean Sea. Nearly all these streams are short in length. Discharges are highest in the wetter northern parts of the country, while in the south perennial flows are rare. The Yarkon, which reaches the Mediterranean just to the north of Tel Aviv, is the largest river of this coastal region. The total stream runoff in Israel is estimated to be 700 to 800 million cubic meters per annum (Kliot 1994, 177).

Two major aquifers exist within Israel. Along the coastal region there is the Plio-Pleistocene or coastal aquifer system. This consists of numerous layers of sands and gravel which were deposited under alluvial fan conditions. Lithologies are very variable, but as an aquifer it functions well. In the highlands of eastern Israel is the Mountain or limestone aquifer system. This is found in Mesozoic sediments, with limestones forming the major water-bearing units.

During the early history of the state of Israel the Coastal Aquifer was seriously depleted as water was withdrawn at a rate of more than double the safe yield (Aberbach and Sellinger 1967). This permitted the rapid settlement of the coastal regions with farming communities, an important part of Zionist ideology. However, in places the water table fell at a rate of a meter a year, and seawater penetrated up to two kilometers inland in the aquifer. This strategy of overpumping the Coastal Aquifer was carried out whilst a major water transport scheme named the National Water Carrier was being constructed (Figure 2.2). The idea behind this was to bring water from the wetter northern part of Israel to the drier south. The source of the water was to be the River Jordan, with the abstraction point located at Lake Kinneret. With the completion of the National Water Carrier in the mid-1960s, major overpumping of the Coastal Aquifer came to an end, and in certain areas the artificial recharge of the coastal plain aquifer commenced.

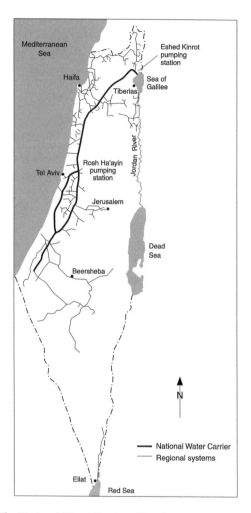

FIGURE 2.2.   The National Water Carrier of Israel.

The late 1960s mark an important turning point in water resource usage in Israel. The total amount of renewable water available to the country is between 1,550 and 1,650 million cubic meters each year. Up until the late 1960s demand had always been less than supply. For example, in 1965 total water use in Israel was only 1,329 million cubic meters. Of this, 80.9 percent was used for agriculture, 15 percent for domestic supply, and 4.1 percent for industry. By the late 1970s annual water demand was 1,700 million cubic meters, and by the mid-1980s it

Million cubic metres

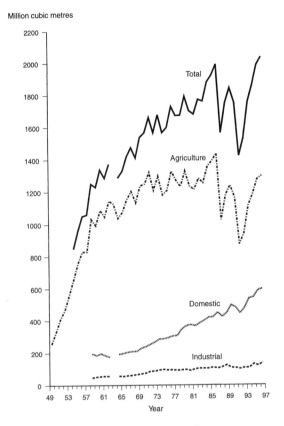

FIGURE 2.3. Changing Patterns of Water Use in Israel.

had risen to almost 2,000 million cubic meters (see Figure 2.3). By this time agriculture alone was using around 1,300 million cubic meters. Interestingly, irrigation in Israel was already utilizing highly efficient methods of application, and so opportunities for further water saving were not great (Beaumont 1993). This high agricultural demand could now only be met by considerable overpumping of the groundwater resources of the nation. Water table levels began to fall again, and it was apparent this time that no other easily developable water resources were available. Although these high levels of supply could be maintained for a number of years by the "mining" of groundwater, it was obvious that they could not be sustained indefinitely. The accumulated deficits in the aquifers are in some cases extremely large. For example, in the Coastal Aquifer the deficit currently attains values of 1,100 to 1,400 million cubic meters and in the Mountain Aquifer 300 to 350 million cubic meters (Kliot 1994, 234).

Israel has always been interested in wastewater reclamation techniques (Beaumont 1993). It is not surprising, therefore, that during the 1970s a large water reclamation scheme was built to the south of Tel Aviv to reclaim the waters of the conurbation. The original aim had been to produce an effluent which was of a standard which would permit its use as drinking water. However, experience showed that such a high standard was not economically feasible and only water suitable for irrigation use could be produced. Unfortunately, reclaimed wastewaters, although a welcome contribution to Israel's water needs, will not solve Israel's water problems, as total volumes remain relatively small. Estimates made in the early 1990s suggest that a total of 485 million cubic meters of water might be reclaimed annually by the year 2000 (Table 2.1). Other authors are even more optimistic, claiming that by 2020 treated effluent flow will reach between 700 and 1,000 million cubic meters, and of this 70–75 percent will be reused (Arlosoroff 1996, 25).

Estimates of water demand for Israel for the year 2000 suggest an annual value of 2,090 million cubic meters (Table 2.1). This, it will be noted, is made up of about 1,600 million cubic meters of water from renewable sources, with an element of mined groundwater, together with a further 485 million cubic meters obtained by water reclamation. It will be noted that although the water is reclaimed largely from urban and industrial sources it can only be used for irrigation purposes. What is really significant about this table is the fact that only 820 million cubic meters of potable water is expected to be used for irrigation purposes. This is in marked contrast to the mid-1980s, when the figure was around 1,200 million cubic meters. If achieved, this will represent a reduction of about one-third in the use of irrigated water. At the same time, the volume of reclaimed water used in agriculture will have increased from 210 million cubic meters in 1985 to 440 million cubic meters in 2000. This is a doubling in less than fifteen years. Perhaps even more significant still is the fact that the total volume of water committed to agriculture will have declined from 1,410 million cubic meters in 1985 to 1,260 million cubic meters in 2000. This represents tangible evidence that the government now recognizes that irrigation cannot continue to use water in such large volumes as it has in the past.

The pattern of actual water use in Israel from the mid-1980s to the

TABLE 2.1. Estimated Water Demand in Israel, A.D. 2000

Sector	Potable Water Million Cubic Meters	Reclaimed Water Million Cubic Meters	Total
Agriculture	820	440	1,260
Domestic	685	—	685
Industry	100	45	145
Total	1,605	485	2,090

SOURCE: Kliot 1994, 241.

present day has revealed startling fluctuations (Figure 2.3). The lowest water usage was recorded in 1991, when the annual consumption fell to a figure of only 1,420 million cubic meters, following a series of dry years in the late 1980s. This value represents a figure similar to that utilized by Israel in the mid- to late 1960s. The main reason for this low figure was a massive cutback in agricultural water use to only 875 million cubic meters. Such a low figure for irrigation had not been recorded in Israel since the mid-1950s. At this time it began to look as if the Israeli public might be willing to accept a much stricter water regime for agriculture. However, the winter of 1991–1992 was one of the wettest on record, and so the imperative to save on water use seemed to become less. Agricultural water consumption began to increase once again, and by 1994 it had reached 1,181 million cubic meters out of a total annual use of 2,019 million cubic meters.

Of particular interest is the way in which domestic usage in Israel has increased steadily over the years. In the mid-1960s it was around 200 million cubic meters each year. By the mid-1980s it was approaching 400 million cubic meters annually, and now in the mid-1990s it is close to 550 million cubic meters. This represents a current annual water use per capita of approximately 100 cubic meters. What does remain surprising is that the industrial water usage in Israel has remained remarkably constant since the early 1970s at around 100 to 130 million cubic meters per annum. This is despite the fact that industry has both expanded and diversified over this long period. The constancy of this figure must lead one to question whether all aspects of industrial water usage are being measured by the published data.

OCCUPIED WEST BANK: A CASE OF WATER NEGLECT

On the West Bank there are few reliable surface water sources, and so underground water has always been important to the local economy. Since its occupation of the West Bank in 1967 Israel has regarded water as a strategic resource to be controlled by military discipline. As a result it has been impossible for the Palestinians to sink wells on their own land without the permission of the military governor. Over the years this permission has rarely been given, and as a consequence the Palestinians have been subjected to increasingly difficult water use conditions as population numbers have grown (Baskin 1993; Wolf 1995). In effect, ever larger numbers of Palestinian people have been forced to survive on what is essentially a fixed volume of water.

From a hydrological point of view the Israeli strategy is quite straightforward. The basic aim of their policy would appear to be to ensure maximum recharge of the Mountain Aquifer in the West Bank so that much of this water will then flow into Israel, where it can be utilized by Israelis. In its simplest terms this could be regarded as a form of "water piracy." What is happening is that the precipitation which falls on the Palestinian West Bank, and which, therefore, could be claimed to belong to the Palestinians, is being encouraged to percolate into the underground aquifer and then flow out of the West Bank toward Israel. At the present time the Palestinians are being prevented from using groundwater so that the maximum amount of water will flow into Israel.

The patterns of water movement in the Mountain Aquifer are relatively well known (Shuval 1992). It is believed that there is a threefold division, with a western aquifer, a northeastern aquifer, and an eastern aquifer (Figure 2.4). The state of Israel has an interest in both the northeastern and the western aquifer systems, as each of them delivers underground water into Israel. The total volumes of water involved are quite considerable, though the exact volumes are still disputed. Shuval (1992) claims that the largest discharge is from the western aquifer, which is believed to have an average flow of around 320 million cubic meters per year. The other two aquifers have discharges of less than half of the western one. The northeastern aquifer has a flow of 140 million cubic meters per annum, while for the eastern aquifer the figure is 125 million cubic meters per annum. In a more recent paper Shuval (1996) presents modified figures for the flows of the Mountain Aquifer. Here it is claimed that the western aquifer has a flow of 310 million cubic meters of fresh water

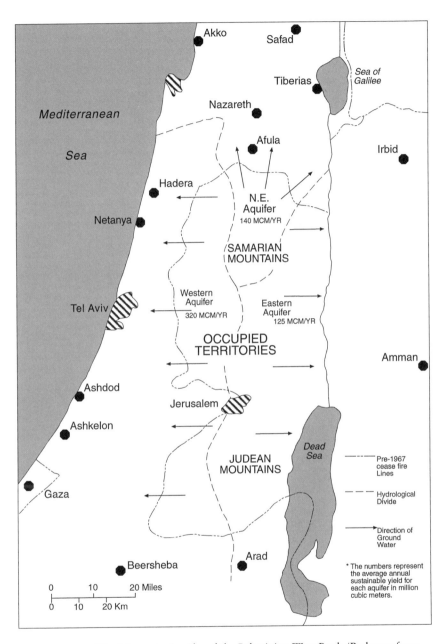

Akko

Safad

Sea of Galilee

Tiberias

Mediterranean

Nazareth

Sea

Afula

Irbid

Hadera

N.E.
Aquifer
140 MCM/YR

Netanya

SAMARIAN
MOUNTAINS

Tel Aviv

Western
Aquifer
320 MCM/YR

Eastern
Aquifer
125 MCM/YR

OCCUPIED
TERRITORIES

Amman

Ashdod

Jerusalem

Ashkelon

Dead
Sea

JUDEAN
MOUNTAINS

Pre-1967
cease fire
Lines

Hydrological
Divide

Gaza

Direction of
Ground
Water

* The numbers represent
the average annual
sustainable yield for
each aquifer in million
cubic meters.

Beersheba

Arad

0   10   20 Miles

0   10   20 Km

FIGURE 2.4. The Mountain Aquifer of the Palestinian West Bank (Redrawn from Shuval 1992).

and a further 40 million cubic meters of brackish water. The fresh water flows of the other two aquifers have been revised downward sharply. The flow of the northeastern basin is now said to be 61 million cubic meters of fresh water and 70 million cubic meters of brackish water. Finally, in the eastern basin fresh water amounts to 81 million cubic meters and brackish water to 70 million cubic meters. The really important point, though, is that the vast majority of the recharge for these three aquifers takes place in the Palestinian West Bank region. From the above it can be seen that Israel would benefit by up to 460 million cubic meters each year if it could maintain sole control over the flow of the western and northeastern aquifers. Its interest in the eastern aquifer is minimal, as this flows away from Israel and down to the River Jordan.

Figures for the actual usage of water on the West Bank are difficult to obtain and are probably manipulated by both the Palestinians and Israelis to strengthen their own points of view. However, estimates from the early 1990s suggest that the Palestinians have access to only about 105 million cubic meters of groundwater, while Israel is making use of approximately 450 million cubic meters (Benvenisti and Gvirtzman 1993; Lonergan and Brooks 1994, 129, 133). Interestingly, an additional 40 to 50 million cubic meters is being extracted from groundwater on the West Bank for new Israeli settlements. In effect, this means that the Palestinians have access only to about 18 percent of the groundwater which is generated on their territories. This provides a measure of the success of Israeli capture of Palestinian water when they are able to utilize about 82 percent of the total amount.

Once the Palestinians gain their independence it is hoped that their access to underground water supplies would increase markedly (Lesch et al. 1992). This would permit the Palestinians to reclaim a portion of the water which is currently draining into Israel. As citizens of an "upstream" country, the Palestinians would be able to argue that they have a right to at least 50 percent of the water which is generated in their territories. In effect, this means that their use of groundwater should be able to rise to a value close to 355 million cubic meters per year. Of this total 160 million cubic meters would be obtained from the western aquifer, 70 million cubic meters from the northeastern aquifer, and a further 125 million cubic meters from the eastern aquifer. With the eastern aquifer the Palestinians would enjoy 100 percent usage, as both the recharge and discharge areas lie within the Occupied Territories of the West Bank.

VALUE OF THE WATERS OF THE JORDAN RIVER

Having examined water use in the individual territorial units, I now focus on water within the basin of the River Jordan itself. Prior to the construction of major irrigation and diversion projects the flow of the Jordan into the Dead Sea averaged somewhere between 1,300 and 1,500 million cubic meters each year. Currently this flow has been reduced to between 250 and 300 million cubic meters (Salameh and Bannayan 1993). It is mostly composed of irrigation return waters from Israel, together with the flow of saline springs. As a result, the water quality is low.

One of the intriguing questions to pose is, what is the value of the waters of the Jordan basin? If it is assumed, for calculation purposes, that the natural flow of the River Jordan is 1,500 million cubic meters each year, it is possible to calculate water values for a series of different uses. The concept of the value of water is a very difficult one to deal with. If water is flowing down a river and it is not used for any human activity it can be claimed to have no value to that particular society. For the moment, any ecological considerations will be ignored. It is only when water is used for economic activity that it gains value. As the water is provided by nature, it can be regarded as a free good, with the only costs associated with it being the costs of supplying the water to the point where it is to be used. This is, after all, how oil is treated. However, the attitude of society toward water is very different from its attitude toward oil. As a result, many water distribution systems have some form of subsidy built into them for certain uses. For example, water used for irrigation is often charged at a much lower rate than that directed to urban/industrial usage, even though the water is taken from the same distribution system. In effect, what has happened is that the cost of the water has been tailored with regard to the consumer's ability to pay for it, rather than reflecting the true costs of water delivery.

This would seem to suggest that one approach would be to consider the value of water in terms of what people would pay to use it for a specific purpose. This then leads on to the issue of who owns the water in the first place and who is permitted to make the decisions with regard to any form of pricing policy. In most countries of the world a form of state ownership of water resources is practiced, although there are considerable variations in the manner in which these ownership rights are applied. In Israel, for example, water resources have been nationalized and are the property of the state.

Of all the possible uses of water perhaps the most valuable one is to sell it as bottled water. In British supermarkets the price of the more expensive bottled French waters is around £1/liter, or US$1.50/liter. This gives a price of US$1,500 per cubic meter. If it were possible to bottle all the flow of the River Jordan and sell it at these prices it would have the quite staggering value of US$2,250,000 million each year. Obviously such a large volume of water could not be sold at these high prices, given the size of the world market for bottled waters. However, the calculation is a useful exercise, as it sets a top limit on the Jordan water's value.

A more realistic use of the waters of the Jordan would be for urban/industrial purposes. Such usage does command a higher volume of demand, but prices are much lower than for bottled waters. In a Welsh context the current cost (1996) of urban/industrial water is £0.7715/cubic meter, or about US$1.15/cubic meter. Using this figure would value the waters of the Jordan River at around US$1,725 million dollars per annum. In itself this is still a very large figure. However, it has to be admitted that at the moment the urban/industrial demand in the countries within the watershed would not sustain all the water being used for such high-value purposes. That being the case, this figure probably represents too high a valuation of the available water at the present time. This position will of necessity change over the next fifty years with the expected population growth.

The final use of water which needs to be considered is that of irrigation. This is a very high-volume use, but costs have to be low, as the crop produced has then to be sold in the marketplace. This means that the price of irrigation water cannot rise above about 20 US cents a cubic meter if the farmers are still to make a profit. Indeed, with many crops the price of the water has to be well below this figure if the farmer is to survive. For the sake of argument let us use a water charge figure of 10 US cents per cubic meter for calculation purposes. This would value the waters of the Jordan at around US$150 million each year. Currently all of the waters of the river could be utilized for irrigation purposes.

In order to assess the value of the waters of the River Jordan it would seem reasonable to estimate that about one-third of the waters could be sold at an urban/industrial rate and the remainder at the irrigation rate. Summing these two figures provides a theoretical value for the waters of the Jordan of about US$669.25 million. What this exercise has shown is that the waters of the Jordan River, however they are used, do provide an important source of wealth for the riparian countries of the basin. In the future a growth of urban/industrial demand is the most realistic usage of water for generating the highest level of economic activity within the region.

It is also worth considering what would be the result if the waters of the Jordan River had to be replaced from another source. The most obvious source would be the desalination of seawater. Current costs for desalination are around US$2 per cubic meter, and so the replacement costs of the Jordan water would be about US$3,000 million per annum. This is a very large figure and is approximately 4.5 times the estimated value of the waters of the Jordan as they are being used currently. It clearly illustrates that as long as the value of current water use in the Jordan basin is so relatively low, there is little incentive for the countries of the region to develop large-scale desalination facilities.

The value of considering water in terms of its replacement costs is that it focuses attention on low-value uses, such as irrigation. It poses the question as to whether these low-value uses can be allowed to continue. The point has to be made, though, that it is only sensible to divert water away from irrigation usage if a demand already exists from a more valuable water use sector. While it is possible to talk in theoretical terms in the case of the Jordan River, political realities that further complicate the situation have to be recognized. The problem here is that the economies of the different entities sharing the basin have very different characteristics. On the one hand, Israel is now beginning to gain the infrastructure of an industrialized state, while on the other, the Palestinian West Bank is still highly dependent on primary production based on agriculture, and industrial activity is at a very low level. Given these differences, the patterns of water demand and water use will be different also.

For the Israelis, the withdrawal of water from irrigation use so that it can be fed into the urban/industrial complex makes good sense, as it permits the generation of greater wealth. However, for the Palestinians this is not really an option, as the urban/industrial complex is not sufficiently developed to make use of the available water. As a result, water will continue to be used by the Palestinians for irrigation in an attempt to generate the wealth which will permit the urban/industrial sector to develop. In effect, what must happen in the Jordan basin is that a series of different water economies will have to come into existence based on the existing territorial realities. In each of these areas water will have different values and uses, some of which may appear perverse from an adjacent and different viewpoint. However, if each of the territorial entities is to further develop its economy it is essential that these water value anachronisms be permitted to exist, at least in the short term. This would inevitably mean that on the Palestinian West Bank irrigation would be regarded as the priority water use, whilst across the border in Israel irrigation water usage would be strongly discouraged. However, selling these ideas to the peoples of the basin will be a difficult and delicate task for the politicians.

FUTURE WATER STRATEGIES: CONFLICT OR COOPERATION

The question of whether a country is willing to go to war over water, or indeed anything else, is a difficult one to answer with any degree of conviction (Beaumont 1997a). There seems little doubt that Israel did not go to war with the Arabs in 1967, 1973, or 1982 with water as the main aim of its strategy. However, it is certain that the Israeli government has always been well briefed on the strategic implications of all resources, including water, in the region. The significance of holding certain land areas as a means to control water resources would have been widely recognized and acted upon.

For many years now the country of Jordan has not been able to utilize any of the waters of the River Jordan below its exit from Lake Kinneret, owing to the very poor quality of the waters released by the Israelis. Indeed, all of the flow of the upper Jordan is currently being utilized by Israel. Much of this water is transported to the coastal region of Israel via the National Water Carrier, with the rest used in the vicinity of Lake Kinneret. This means that the only flows which occur along the mainstream of the Jordan are irrigation return waters, the flows of saline springs, and occasional floodwaters. As a riparian state, Jordan obviously has a claim to at least some of the waters of the upper Jordan, but it will be difficult, now that a peace treaty has been signed, to obtain a greater share of the waters than the miserly amount it has been allocated (Beaumont 1997b).

To date all the talk has been of swapping land for peace, but the swapping of water is probably even more important. The key point in this discussion, therefore, is whether Israel would be willing to relinquish, each year, about 200–250 million cubic meters of water from the Jordan River, and a further 230 million cubic meters of groundwater from the Mountain Aquifer, to secure peace. The signals from the Israelis have not been auspicious. Israeli commentators on water matters almost always imply that the solution to the region's water problems is to obtain extra water from elsewhere. The most-cited examples are water from the Litani River in Lebanon, the River Yarmuk, or the River Nile in Egypt. For example, Kally states: "The practical solution to the problem is to import water into the Territories from external sources, such as the Yarmuk, the Nile, and the Litani rivers" (Kally 1993, 53). Similarly, Shuval (1996) states: "Syria and Lebanon are the only two of the five Jordan riparians that will almost surely be able to meet their own MWR [minimum water requirements] and are logical potential candidates to assist their Arab neighbors." Almost no consideration seems to be given to a more equitable distribution of the existing resources within the region. Indeed, Kally (1993) proposes giving "Israeli" water to the Palestinian West Bank in exchange for Nile water for the Negev. What he does not mention is the fact that the so-called "Israeli" water is really groundwater generated on the West Bank in the first place which has been expropriated by Israeli actions. In effect, what is being suggested is that the Palestinians should be given their own water back if the Israelis can get access to new water from the Nile! The need for future cooperation in the face of growing water shortages is recognized by at least one Israeli author: "There is already a water crisis in certain areas, notably Jordan and Gaza, which is likely to spread into the West Bank and Israel. Only cooperation can alleviate it" (Fishelson in Kally 1993, 24). Unfortunately, the nature of the cooperation does not appear to include returning water to the Arabs.

If true cooperation is to occur between the Israelis and the Arabs in the future, then Israel must accept that for over thirty years its economy has benefited from the use of water which many people would claim does not belong to it. Israel's economy is now so mature that the use of large volumes of water for irrigation is no longer necessary. Today agriculture only employs about 3 percent of the civilian labor force, yet still utilizes 58 percent of the nation's water resources (1994). Israel must take the lead and rapidly run down its agricultural water consumption. This will free the water which is needed to give back to the Palestinians and the other Arab states to permit their economies to develop.

In the long term, it seems inevitable that Israel first, and Jordan soon afterward, will become increasingly dependent on water supplies other than those which are already available to them. Any hope of solving the water problems of the Jordan basin by importing water from the Litani River or the Nile River seems rather short-sighted, as these water resources will be needed by both Lebanon and Egypt within the next decade or two. It would seem far better to accept this fact and to concentrate on new sources of supply. For the Israelis the obvious solution would be the desalination of seawater. The technology is now well proven and costs are dropping to a level where the Israeli urban regions could absorb the higher water costs without great hardship.

If a state of genuine cooperation existed between the Israelis and the Arabs it would make considerable sense to devise joint projects to supply a series of urban regions on either side of the political divide. For example, a single desalination plant could supply both Aqaba and Eilat, while facilities on the Mediterranean coast of Israel could provide water for Israeli cities as well as the larger urban centers of the Palestinian West Bank. This is not to suggest that a single water economy could come into existence, as the different states of economic development dividing the countries, mentioned earlier, have to be accommodated. For a number of years the Palestinians would still require access to cheap irrigation water if there was to be any hope of their economy flourishing and maturing.

The biggest challenge undoubtedly faces the Israelis, as it is their military might which dictates the realities on the ground. They have the skills and technology to carry out an ambitious program of restructuring water use within their own country and also the experience to help the Arab nations develop their own economies through more efficient water use. Whether the stability and prosperity of a lasting peace are sufficient to make the Israelis overcome their distrust of their Arab neighbors to such an extent that they will give back the water they have expropriated will be a test of the leadership of present and future Israeli governments. Fortunately for all, the conflict stage amongst the nations of the Jordan basin now seems to be over forever. The present coexistence, although peaceful, is still capable of generating considerable tension and unrest. Genuine cooperation still remains a future objective, but it could be through the agreed reallocation of the waters of the Jordan that it is achieved.

REFERENCES

Aberbach, S. H., and A. Sellinger. 1967. "Review of Artificial Recharge of the Coastal Plain of Israel." *Bulletin of the International Association of Scientific Hydrology* 22: 65–77.

Allan, J. A., ed. 1996. *Water, Peace and the Middle East: Negotiating Resources in the Jordan Basin.* London: Tauris Academic Studies.

Arlosoroff, S. 1996. "Managing Scarce Water: Recent Israeli Experience." In J. A. Allan (ed.), *Water, Peace and the Middle East: Negotiating Resources in the Jordan Basin,* pp. 21–48. London: Tauris Academic Studies.

Baskin, G. 1993. "The West Bank and Israel's Water Crisis." In G. Baskin (ed.), *Water: Conflict or Cooperation*, special issue, *Israel/Palestine Issues in Conflict: Issues for Cooperation* 2 (2): 1–11. Jerusalem: Israel/Palestine Center for Research and Information.

Beaumont, P. 1989. "Wheat Production and the Growing Food Crisis in the Middle East." *Food Policy* 14 (4): 378–384.

———. 1993. *Drylands: Environmental Management and Development.* London: Routledge.

———. 1994. "The Myth of Water Wars and the Future of Irrigated Agriculture in the Middle East." *International Journal of Water Resources Research* 10 (1): 9–21.

———. 1997a. "Water and Armed Conflict in the Middle East: Fantasy or Reality?" In N. P. Gleditsch (ed.), *Conflict and the Environment*, pp. 355–374. Symposium of International Peace Research Institute. Oslo: Dordrecht, Kluwer.

———. 1997b. "Dividing the Waters of the River Jordan: An Analysis of the 1994 Israel-Jordan Peace Treaty." *International Journal of Water Resources Development* 13 (3): 415–424.

———, G. H. Blake, and J. M. Wagstaff. 1986. *The Middle East: A Geographical Study.* London: David Fulton Publishers.

Benvenisti, E., and H. Gvirtzman. 1993. "Harnessing International Law to Determine Israeli-Palestinian Water Rights: The Mountain Aquifer." *Natural Resources Journal* 33: 544–567.

Haddadin, M. J. 1996. "Water Management: A Jordanian Viewpoint." In J. A. Allan (ed.), *Water, Peace and the Middle East: Negotiating Resources in the Jordan Basin*, pp. 59–73. London: Tauris Academic Studies.

Kally, E. 1986. *A Middle East Water Plan under Peace.* Tel Aviv: Armand Hammer Fund for Economic Co-operation in the Middle East.

———, with G. Fishelson. 1993. *Water and Peace: Water Resources and the Arab-Israeli Peace Process.* Westport, Conn.: Praeger Publishers.

Kliot, N. 1994. *Water Resources and Conflict in the Middle East.* London: Routledge.

Lesch, A. M., et al. 1992. *Transition to Palestinian Self-Government: Practical Steps towards Israeli-Palestinian Peace.* Report of a Study Group of the American Academy of Arts and Sciences. Bloomington: Indiana University Press.

Lonergan, S. C., and D. B. Brooks. 1994. *Watershed: The Role of Fresh Water in the Israeli-Palestinian Conflict.* Ottawa: International Development Research Center.

Lowdermilk, Walter Clay. 1944. *Palestine: Land of Promise.* London: Gollancz.

Lowi, M. R. 1993. *Water and Power: The Politics of a Scarce Resource in the Jordan River Basin.* Cambridge: Cambridge University Press.

Population Reference Bureau. 1995. "1995 World Population Data Sheet." Washington, D.C.: Population Reference Bureau.

Salameh, E., and H. Bannayan. 1993. *Water Resources of Jordan: Present Status and Future Potentials.* Amman, Jordan: Friedrich Ebert Stiftung.

Shahin, M. 1989. "Review and Assessment of Water Resources in the Arab World." *Water International* 14: 206–219.

Shuval, H. 1992. "Approaches to Resolving the Water Conflicts between Israel and Her Neighbors: A Regional Water-for-Peace Plan." *Water International* 17 (3): 133–143.

———. 1996. "Towards Resolving Conflicts over Water between Israel and Its Neighbors: The Israeli-Palestinian Shared Use of the Mountain Aquifer as a Case Study." In J. A. Allan (ed.), *Water, Peace and the Middle East: Negotiating Resources in the Jordan Basin*, pp. 137–168. London: Tauris Academic Studies.

Wolf, A. T. 1995. *Hydropolitics along the Jordan River: Scarce Water and Its Impact on the Arab-Israeli Conflict.* Tokyo: United Nations University Press.

3. *Forces of Change and the Conflict over Water in the Jordan River Basin*

Steve Lonergan

INTRODUCTION

The bombing of Lebanese power plants by Israel in April 1996, which effectively cut off power to most of the country, underscored the continued relationship between resources and armed conflict in the Middle East. The use of resources as "strategic targets" is, of course, not of recent origin, nor is it confined to only one state. Modification of resources and the environment for military purposes is a long-standing issue. In 146 B.C., the Romans plowed salt into the farm fields around Carthage, destroying the city's economic base; prior to this, in about 2400 B.C., Sumerians dug a canal to divert water from the Tigris to the Euphrates to gain independence from Umma (Roots 1992). Energy production and transmission facilities and water supply projects are often primary targets of military activity, and another recent example is the destruction of Kuwaiti desalination plants by the Iraqis during the Gulf War.

Much has been written about water in the Middle East, and the purpose of this chapter is not to simply repeat that information (cf. Lonergan and Brooks 1994; Kliot 1994; Hillel 1995; Wolf 1995). What is important is to recognize that the role of water—and other resources—in the region is historically, spatially, and socially constructed; what appears to be a simple link between water resource scarcity and conflict is mired in a complexity of issues. Therefore, decisions involving water—as well as agreements among parties in the region—have social, economic, and *spatial* implications.

Accordingly, the purposes of this chapter are threefold. First, to reiterate that the relationship between water scarcity and conflict in the Middle East cannot be reduced to a simple cause and effect relationship, but is merely part of a very complex system of linkages. Conversely, the use of water has been a subtle, but underlying, factor in exacerbating the conflict between Israel and the Palestinians (and, indeed, among most neighboring groups or states in the region). Second, to address whether the forces of global change—population growth, urbanization, and global warming in particular—will have a major effect on the potential for conflict over water in the future in this region. And last, to focus on the issues of whether water rights are a major barrier to peace in the region and whether the development of water markets might be a solution. Throughout the chapter there is a concern for the spatial aspects of these issues.

THE LINKS BETWEEN WATER SCARCITY AND SECURITY IN THE MIDDLE EAST

Water problems in the Middle East—and particularly in Israel, Jordan, and the West Bank and Gaza Strip—are not limited simply to considerations of scarcity, but are a result of three interrelated and interacting crises (Lonergan and Brooks 1994). The first crisis is one of water supply and demand. Since the mid-1970s, demand has outstripped supply. Population growth through natural increase and in-migration places greater demands on the provision of water for domestic consumption, and many proposals for economic expansion will likely increase demand further (see Figure 3.1). Yet there is no evident way to increase supply.

The second crisis involves deteriorating water quality. In Israel, Jordan, and the West Bank–Gaza, water has been—and is still being—polluted by growing volumes of industrial and agricultural wastes, and in some cases by human sewage. A recent report by the Institute of Advanced Strategic and Political Studies in Israel noted that the quality of drinking water in Israel is below U.S. standards of thirty years ago. "Don't drink the water in Israel," was the report's conclusion (*Jewish Post and Opinion*, 22 May 1996).

The third crisis relates to geopolitics. Roughly one-third of the water consumed in Israel comes from groundwater that originates as rainfall over the West Bank—on land that in any final settlement is likely to belong to Palestinians. In addition, 50 percent of the flow of the upper Jordan originates in Lebanon and the Golan Heights. Recent statements by the Netanyahu government have made it clear that both security and water resources in the West Bank will continue to be controlled by Israel. Even if the issue of water allocations between Israel and the Palestinians is settled, a major conflict between the two groups is emerging over the issue of water rights. The relationship among Israel, Jordan, and Syria is no less tenuous. Although Jordan and Israel have reached an agreement

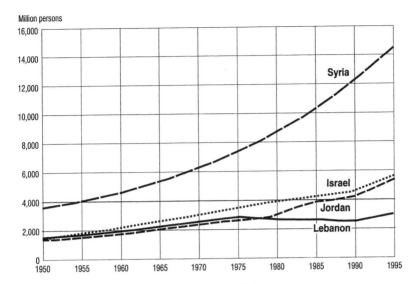

Million persons

FIGURE 3.1. Population Growth: Selected Middle East Countries, 1950–1995.

regarding the Yarmuk River (see the chapter by Kliot later in this volume), the exclusion of Syria from this agreement makes its implementation uncertain.

The first two crises are common to many countries, and even the third crisis reflects conditions that are anything but rare in the region. Bahrain, Kuwait, Libya, Qatar, Saudi Arabia, and the United Arab Emirates are all consuming much more water than their annual renewable water supply, and Egypt, Libya, Oman, and the Sudan are fast approaching the same situation (Rogers 1994). More specifically, whatever the ultimate resolution of the Israeli-Palestinian conflict, both peoples—and the neighboring Arab states—will have to deal in the near term, in the middle term, and in the long term with water quantity and water quality problems. The goal of "sustainable livelihoods" seems inexorably bound up with the need to resolve conflicts over water.

These three "crises" indicate the complexity of water scarcity issues in the region. It is clear that water will remain a "strategic" resource for all states and territories in the region, and for this reason, it has been an important component of the ongoing peace talks. Previous writings by many of the authors in this volume have highlighted the complex nature of this issue. But there remains the question of whether water has been— or will be—a contributor to conflict in the Middle East. Alternatively, could it also be a catalyst for change and for peace in the region?

There is little doubt that there have been tensions over water in the past, whether one is referring to the Nile, the Jordan, the Litani, or the Euphrates. But how have these tensions manifested themselves? An excellent typology of resource-related conflicts is provided by Gleick (1990), and it is a useful framework within which to view this, and other, resource conflicts. He classifies resource and environmental threats to security in five categories:

1.   Resources as strategic goals;

2.   Resources as strategic targets;

3.   Resources as strategic tools;

4.   Resource inequities as roots of conflict; and

5.   The disruption of environmental services as roots of conflict.

Resources as Strategic Goals

An obvious example of considering resources as "goals" to be achieved are territorial conflicts over energy resources. With the advent of greatly expanded international trade and world spot markets for resources, this link between resources and international conflict may be weaker now than in the past, but it remains important in many situations. Setting such goals need not involve direct military activity. The importance of water to a future Jewish state was obvious to many of the early Zionists, and proposals to the Paris Peace Conference in 1919 had the explicit goal of including the Litani and Yarmuk Rivers in such a state. Resources as "goals"—and the potential for conflict—become more relevant when these resources are inequitably distributed over space or through society.

Resources as Strategic Targets

Energy production and transmission facilities, as well as water supply projects, are often primary targets of military activity. Numerous examples are noted in Lonergan and Brooks (1994). Some of the more interesting ones include the following. When the British left Palestine in 1948, Arab groups immediately cut the water supply to Jerusalem. (Water supply to Jerusalem had been severed prior to this, but was repaired by the British.) Jewish residents of the city survived because *it was expected* that the water supply to the city (water was pumped up the

Latrun Valley from 60 km to the west) would be an initial target of Arab groups. Less than two decades later, one of the first activities of the newly formed PLO in 1964 was an attempted sabotage of the Israeli National Water Carrier (Wolf 1995). And a more recent example of resources as strategic targets was the destruction of Kuwaiti desalination plants by the Iraqis in the Gulf War.

Resources as Strategic Tools

The nature of water as a resource which moves across borders leads to the possibility that upstream riparians might use the resource as a tool of control against downstream groups. The most obvious present-day example of this is the threat by Turkey to restrict the flow of the Euphrates to Syria and Iraq in order to pressure Syria to discontinue its support of Kurdish separatists in Turkey. Despite assurances by then–Turkish President Turgut Özal that the country would never hold downstream riparians hostage by restricting the flow of the Euphrates, it is obvious from the specific threat to Syria that Turkey would be willing to use water as a strategic tool. Further examples can be found in Lonergan and Brooks (1994).

Resource Inequities as Roots of Conflict

Considering resources as targets or goals may be linked to a number of root causes, one of which is an inequitable distribution of these resources. Growing disparities in the access to resources between resource-rich and resource-poor regions/groups have created a tension in some areas and open rebellion in others. Such inequities may result more in internal strife than international conflict, and increasingly may be the cause of resource-based conflicts.

Environmental Services and Conditions as Roots of Conflict

The final category relates to the possible disruption of the waste-assimilative capacities of ecosystems or the deliberate manipulation of the *flow* of environmental services. This issue is presently being addressed by the United Nations General Assembly in its proposed revisions to the Convention on the Prohibition of Military or Other Hostile Use of Environmental Modification Techniques (popularly called the ENMOD Convention). The Convention was originally signed in

May 1977, and at that time focused more on weather modifications and modifications of Earth processes such as earthquakes and tsunamis for strategic use. Proposed revisions will expand the Convention to include a broad array of environmental processes, such as forest fires and the massive release of airborne pollutants (for a complete discussion of the ENMOD Convention, see Roots 1992). Deliberately contaminating another's water supply is the most obvious example of disrupting environmental services.

Although these categories are not mutually exclusive, and Gleick fails to identify the links among these categories, they are convenient in characterizing the different uses of resources and the environment for strategic purposes. Most, if not all, of these five "uses" of water can be identified with countries in the Middle East. The spatial implications of considering water a "strategic" issue—at least for the Palestinians—are clear. Agricultural expansion is severely constrained; nonessential uses of water are either foreclosed by the military authority or limited due to discriminatory pricing policies; and infrastructure development is subject to Israeli authority. Such spatial outcomes are the direct result of limited Palestinian access to, and control over, water supplies.

Other examples relate to upstream control and use of water resources in the region. Turkey has already affected the flow of the Euphrates during the construction of its Southeast Anatolia Project; Syria has reduced the flow of the Yarmuk with its upstream dams; and Israel has polluted the Jordan River through its diversion of saline springs from Lake Kinneret into the river. Most—if not all—of these activities are also conditional on military superiority: recognizing this and taking strong exception, Egypt has already given notice to upstream nations that it will not tolerate any reduction in the flow of the Nile due to increased withdrawals.

It is clear from these few examples—and the many other writings on the issue in this book and elsewhere—that water is, indeed, a strategic resource in this region. Given this fact, how might social, economic, and environmental changes in the future affect water scarcity and, in turn, the potential for conflict over water?

GLOBAL CHANGE AND WATER RESOURCES

This section addresses three issues with respect to water scarcity internationally and, in particular, in the Middle East: population growth, urbanization, and climate warming.

Population Growth

Figure 3.2 depicts water consumption and water availability for selected countries in the world in 1995 and 2025 (using the medium UN population forecasts for 2025). Under the UN's high-population scenario, by the year 2050, two-thirds of the world's population could be experiencing conditions of water stress or water scarcity (Figure 3.3). Water stress, or the water stress index, is based on countries which have renewable freshwater resources less than 1,700 cubic meters per capita per year. Below 1,000 cubic meters per capita, most countries are likely to experience chronic water scarcity on a scale which is sufficient to impede economic development and harm human health. However, it should be noted that these are merely "benchmark" levels. Countries will respond differently to conditions of water stress and scarcity depending on whether they are energy-rich, like Saudi Arabia, and therefore can afford large desalination projects; whether they are wealthy, like Singapore, and can purchase water from neighboring water-rich states (Malaysia); the efficiency with which they use water; and the water intensity of their economies. To make matters worse, there are indications that per capita consumption will not remain constant, but will actually *increase* in many of these countries. This is due to two factors. First, the level of develop-

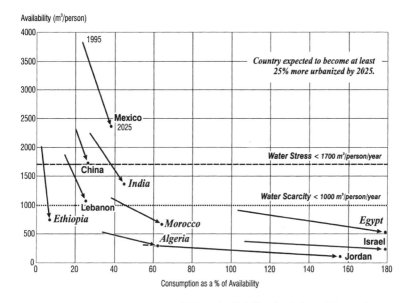

FIGURE 3.2. Water Consumption/Water Availability for Selected Countries, 1995 and 2025.

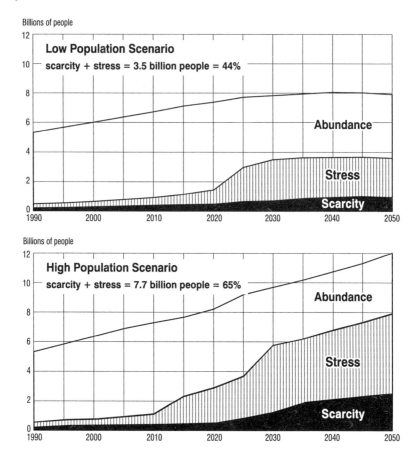

FIGURE 3.3. Population Experiencing Freshwater Scarcity, 1990–2050.

ment, since resource consumption, including water consumption, is positively related to levels of economic development. This is true even with the transfer of water from agricultural uses to industrial/commercial uses. And second, increasing urbanization will also act to increase water consumption per capita.

Water Problems in Urban Areas

In one of the opening sessions at the United Nations Habitat II Conference in Istanbul in early June 1996, Wally N'Dow, the Secretary General of the Conference, told the participants, "The scarcity of water is replacing oil as a flashpoint for conflict between nations in an increasingly urbanized world." He goes on to say that "water stands today as one of the most critical dangers, one of the most critical breakdowns of peace between nations. It has replaced the threat of war over oil." While this may be overstating the problem, it is true that rapid urbanization in many regions of the world, along with decaying infrastructure, is placing severe strains on systems of water supply and wastewater management. Water shortages are becoming a major problem in the world's megacities (cities with more than 10 million people). The extreme water shortages faced by residents of cities in the developing world will threaten their lives and health. The three main causes of the impending urban water crisis were noted as:

1. Rapid urban population growth, increasing at the rate of 170,000 persons per day in developing countries;

2. Fifty percent of all potable water being wasted or lost in the developing world; and

3. Pollution, with over 2 million tons of human excrement and an ever-increasing volume of untreated discharge going into urban water supplies every day.

In addition, in the developing world, more than 1 billion people do not have access to clean drinking water, and 1.7 billion people lack access to adequate sanitation facilities. The UN goes on to note that dirty water causes 80 percent (or more, according to the World Resources Institute 1996) of diseases in the developing world and kills 10 million people annually.

This indicates—potentially—a crisis of epidemic proportions. And while urban water use amounts to less than one-tenth of total water use on a global scale, urbanization increases the per capita demand for water for domestic purposes. This is the result of three forces. First, the rise in industrial water demand. Second, better access to water results in higher demand. And third, higher numbers of people require greater amounts of food, and irrigated agriculture near cities tends to increase.

Two other issues come into play as well. First, water is generally priced much lower than its marginal cost and, therefore, the incentive is not to conserve water. And second, unaccounted-for water comprises a very large portion of total water supply. Unaccounted-for water is simply water that is "lost" in the system, usually by one of three means. Decaying infrastructure with leaky pipes (in some cases, such as in Bangkok and refugee camps in the West Bank and Gaza, sewage actually seeps into the water supply system). Faulty meters are a second problem. And so-called "black water," or water that is withdrawn using illegal connections, is a third source of this unaccounted-for water.

A final problem relating to urban growth and water availability rates mention. In certain cities which are very dependent on underground sources of water, the water is being pumped out faster than it can be naturally replenished. This has resulted in two problems. First, cities such as Bangkok and Mexico City are experiencing subsidence. In Mexico City, the groundwater level is sinking by approximately 1 meter per year (World Resources Institute 1996), and the Mexico City Metropolitan Area has fallen by an average of 7.5 meters over the past century. In Bangkok, the situation is similar, as overdrawing the aquifer has led to land subsidence on the order of 10 centimeters per year in some parts of the city.

In coastal regions, whether the Gaza Strip, Israel, or coastal Florida, overpumping aquifers has led to saltwater intrusion into the aquifer. The Coastal Aquifer in Gaza, for example, already has levels for chlorine salts which are three times the World Health Organization (WHO)–recommended levels.

Will these two forces—increased population levels and increased urbanization—affect the situation with respect to water in the Middle East? The answer is yes, but less so than in other regions. Israel is already heavily urbanized (over 90 percent of Israelis live in urban areas; World Resources Institute 1996), but continued in-migration will be concentrated in urban areas and, therefore, affect the demand for water. Some of this demand could be met by the increased use of recycled water in the agricultural sector. A larger problem looms in countries such as Jordan, where the population of Amman is growing at 6 percent annually. It is too simplistic to assume that more efficient use of water in agriculture (through improved technology or higher prices, for example) will offset the growth in demand for water in urban areas. Sources of water may be distant from cities; increased sewage production affects water quality; and increased demand for infrastructure (for both water supply and wastewater removal) is very costly. However, it is fair to say that one of the only obvious sources of additional fresh water for domestic use is water presently used by the agricultural sector. The spatial implications of this are clear: a shift to less water-intensive crops, conversion of marginal agricultural lands to other forms of production, and increased use of recycled water. It appears, then, that increased population growth and increased urbanization—both within the region and globally (see Figure 3.4)—will significantly reduce the amount of water available per capita

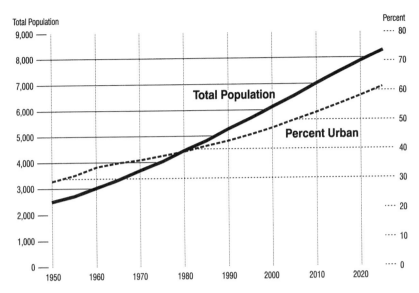

FIGURE 3.4. World Total Population and Percent Urban, 1950–2025.

in many regions of the world, including certain regions in the Middle East; this will, undoubtedly, increase social tensions and could continue to be a contributor to conflict in some of these regions.

Climate Change

One of the great uncertainties with respect to water in the Middle East is how climate change might affect the availability of water in the region, and whether steps should be taken now to incorporate climate change projections into water resource development planning. Studies of the impacts of climate change on water resources in the Middle East (see, for example, Lonergan and Kavanagh 1991) indicate that increased concentrations of greenhouse gases in the atmosphere will likely result in a general increase in temperature in the region. Projections of precipitation, however, are inconsistent; some of the general circulation models (GCMs), which project future climate, show an increase in precipitation, while others show a decrease. However, the models also indicate greater variability in the magnitude and frequency of extreme events, such as droughts.

One important factor to remember about precipitation in the region—and particularly in the Jordan River Valley—is its seasonality. Virtually no rain falls in the summer months; over 95 percent falls between November and April. In some situations, the problem of water scarcity is not simply one of absolute scarcity, but relative scarcity. Ample long-term storage systems could go a long way toward reducing the problems of water scarcity in the region.

Will these projected changes in temperature and precipitation have a significant impact on the availability of water in the region? The key will be a combination of changes in precipitation and changes in evapotranspiration. As Turkey continues with its development of the Southeast Anatolia Project (GAP), the enormous surface water reservoirs which are part of the project will be affected by increases in surface water evaporation. Increases in evapotranspiration could be as high as 20 percent per year, depending on which GCM one believes (see Lonergan and Kavanagh 1991). The implications are clear: less surface water, likely less rainfall and recharge to aquifers, lower soil moisture, and higher water demand (for crops and humans).

Three items are worthy of note at this point. First, these projections are based on the output of macro-level models which are not able to take into account micro-climatic variations. The tremendous spatial variation in temperature and precipitation in this region, and the seasonality of rainfall, make the broad generalizations about changing climate very speculative. Second, the variability in the projections of changes in precipitation across models is a concern. Although there is reasonable consistency with temperature projections, those of precipitation are much more variable. And last, while it is important not to trivialize the potential role climate change may play in affecting water availability in the region, rapid population growth, inefficient use of resources, decaying infrastructure, deteriorating water quality, and the lack of a comprehensive regional water plan are much more crucial in terms of the potential for water to contribute to conflict in the future.

What will the spatial effects of increases in temperature and rates of evapotranspiration be in smaller regions such as the Jordan Valley? The general move away from water-intensive crops such as citrus and bananas will minimize the impacts of a changing climate. But ultimately, there will need to be a shift away from agriculture to less water-intensive activities. Even a move to rain-fed agriculture could prove problematic in an era of higher temperatures and lower rates of precipitation. Development of surface water storage must also take into account possible changes in climate. And while the impact of climate change for most residents in the Jordan Valley will be minimal, even the marginal changes in water availability brought about by lower precipitation rates mandate a move to more appropriate pricing of water, where possible.

RESOLVING WATER CONFLICTS
WITH WATER MARKETS

The issue of water pricing as a mechanism to promote water conservation has been discussed in much detail over the past few years. The World Bank has been a strong advocate of higher water prices in countries, such as Jordan, where water is heavily subsidized (World Bank 1993). As part of this move to "market-oriented" instruments to promote water conservation, a number of authors have recommended the establishment of water markets as a means of overcoming the difficult problems pertaining to water allocations between Israel and the Palestinians. "Water markets" simply refers to a situation where the forces of supply and demand determine the price of the good: in this case, water. Individuals or organizations would be allowed to bid for water until the price is determined by how much people are willing to buy and how much they are willing to spend. If we accept that water is a scarce resource in the region, why not develop water markets as a mechanism which will promote efficiency of use? The answer to this question is that water markets will not be an effective mechanism in this region for one simple reason: they won't work. And this is partly for practical reasons and partly for what we might call "symbolic" reasons.

Kenneth Boulding's (1964) perspective on water is appropriate in this context. He notes that "The perception of water as a symbol of ritual purity exempts it to a certain extent from the dirty rationalities of the market." The criticisms of developing efficient allocation systems for water—through water markets, for example—in the Jordan Valley, however, are based on more specific criteria than Boulding's notion of the symbolic importance of water. They include (after Baumol and Oates 1975):

1. The focus on efficiency of use ignores principles of equity. The market simply assumes that all potential buyers are equally well-off.

2. The market does not, necessarily, reflect needs as much as it does wants.

3. Efficiency criteria also ignore future generations. The efficient allocation of resources is efficient only for today's generation; future generations have no influence in decisions, and must be looked after by the State (which, in turn, must affect the operation of the market).

4. Externalities—or external factors of production, which may be positive or negative—are not included in the search for efficient re-

source allocations. These externalities can include the social costs of pumping an aquifer or discharging wastes into a stream, for example.

5. For a market to work effectively, property rights must be well defined. The extreme example of this is the difficulty in valuing collective or "public" goods.

The issue of equity alone makes the use of water markets untenable. The per capita income of Israelis is ten times that of Palestinians in the West Bank and Gaza. The notion that Israelis could buy water from Palestinians (assuming an equitable allocation of water occurs), with the proceeds of these sales then being reinvested to produce jobs for Palestinians, is almost absurd. These are two economies in very different stages of development, and assuming there is equity over space is simply not true.

It is also the case in most jurisdictions that property rights for water are either poorly defined or are mixed among individuals, the State, and the public. At present, "rights" over water throughout Israel (see the chapter by Kliot in this volume for current developments in water rights in Israel), the West Bank, and Gaza lie primarily with the State of Israel, although there are some privately owned wells and municipal wells in the West Bank and Gaza. Immediately, this poses a problem for the operation of water markets. In addition, it affects another requirement of water markets as well: that there be many buyers and sellers. However, markets can still work well when water can be allocated to individuals. Wishart (1993) has called for the establishment of a water "bank" in Israel, where water could be bought and sold, and feels that the water distribution system in the country, as well as the ability of the State to allocate water to individuals, would ensure a strong market for water. This assumes, however, that the distributional costs of delivering water are zero (or averaged over the entire country). The situation is less clear for the interjurisdictional situation with groundwater, however, as the issue of property rights remains muddled.

A further requirement relates to the mobility of water. For markets to operate effectively, there needs to be perfect resource mobility. In the Jordan Valley, there is a political as well as a physical aspect to water mobility. While Mekorot has a well-developed water supply system, it is unclear whether the distribution of water to the Palestinians by Mekorot is an acceptable situation. Such an arrangement has immense implications for Palestinian sovereignty and will undoubtedly be an important aspect of the final status negotiations between Israel and the Palestinians. It can be said that at least the physical infrastructure needed to promote water trading is available; this may be appropriate in the short term until a Palestinian-operated system becomes available.

So, are the conditions appropriate for the development of water markets between the Israelis and the Palestinians? The lack of well-defined property rights for water in the region poses an enormous barrier to the development of water markets. Until water allocations, a division of water, or water rights have been decided on, it would be difficult to imagine a smoothly operating market for water. More importantly, the key issue to a peaceful future for Israelis and Palestinians is addressing the many inequities which exist between the two groups. Since water markets focus specifically on principles of efficiency and ignore equity issues, they are inappropriate for the foreseeable future.

THE TABA AGREEMENT

On 28 September 1995, Israel and the Palestinians signed the Taba Interim Agreement ("Oslo II") as part of the peace process which began in Madrid in 1991. The Interim Agreement contained provisions for future relations between Israel and the Palestinians, and called for the establishment of a Palestinian Council and permanent status negotiations (which would begin by May 1996). One of the key provisions (Article 40) of the agreement related to water and sewage. Some of the key elements of this article include the following:

1. A recognition of Palestinian water rights in the West Bank. However, these "rights" were not defined and will be negotiated in the permanent status talks. What is clear is that Israel interprets these as rights to *amounts* of water, while the Palestinians interpret these as rights to *control* over water.

2. The parties agreed to cooperate on the joint management of water and sewage systems in the West Bank. This cooperation includes joint management and development of additional water resources as well as sharing all relevant data—including extraction rates, consumption data, and so on.

3. A joint water management committee (JWC) will be established whose responsibilities include: coordinated management of water resources and water and sewage systems; protection of water resources; exchange of information; overseeing supervision and en-

forcement mechanisms; the resolution of disputes; water purchases from one side to another; and measurement and monitoring.

4. Cooperation in water-related technology transfer, research and development, and the development of future sources of supply.

5. Acceptance of the future needs of Palestinians in the West Bank of 70–80 mcm per year, and the immediate transfer of 28.6 mcm per year.

Despite the apparent force of the provisions of the Taba Agreement, many of the sections are vague and contradictory. Any contentious areas—such as water rights—will be further negotiated in final status talks. The agreement also includes general statements about the need to maintain existing utilization rates and recognizing the need to develop new sources. With respect to water, it is a very weak agreement for the Palestinians. However, the election of Benjamin Netanyahu as prime minister of Israel further altered this situation. While it is likely that the Palestinians will be allowed to run many of their own affairs on the West Bank, it is also clear that foreign affairs, security, and *water resources* will continue to be controlled by Israel. Thus, the provision of the agreement which states that "powers currently held by the civil administration and military government relating to water and sewage will be transferred to the Palestinians" has virtually been nullified under the pronouncements of the new administration. In practice, there has been no transfer of authority to the Palestinians, no new wells have been agreed upon, and severe water shortages remain (Isaac and Kubursi 1996).

The effects of the Taba Agreement on water use and the Palestinian economy in general have been virtually nil. As was noted above, it is clear that water resources remain a strategic and security concern for Israel, and, as such, they will remain firmly under Israeli control, at least for the near future.

CONCLUSION

The discussion sought to provide insights into two very different perspectives on the problem of water scarcity in the Middle East in general, and the Jordan River Valley more specifically. First, whether "global" changes in population, urbanization, and climate might affect the water situation in the Middle East; and second, whether water markets, which have been much promoted as a solution to water problems in the region, would be a viable mechanism given the historical and cultural context of water resource use and development. The chapter ended by summarizing the recent agreement between Israel and the Palestinians, and considering whether this agreement addresses any of the key issues raised in earlier sections of the chapter. Together, these represent three "forces of change" presently acting on, and in, the region. Environmental change, both through the natural environment and via strong social trends such as urbanization; economic change, with the pressure for higher-priced water and the need for economic development; and institutional change, through bilateral and multilateral agreements which affect the utilization of water. Any discussion of the importance of water in the Middle East must not only be set in its social, economic, and historical context, but must incorporate these forces of change as well.

It is clear that many other global forces affecting society will have an impact on the Middle East as well, including population growth and movement, conflicts which might emerge between culture and development, and the expansion of trading blocs. However, it is apparent that in the present context, the concern for these forces takes a back seat to other more pressing problems when dealing with water. While this is understandable, it is also myopic, since only with major structural changes can the problems of water scarcity in the region be resolved.

With respect to the external pressure for higher water prices and the feasibility of water markets, concerns with distributional issues—relating to unequal economic power between institutions or regions and between generations, the pervasiveness of externalities, and the problem of public goods—mandate that some measure of government interference is necessary in dealing with water management. Water is not simply an economic good; it is also essential for human survival and for the maintenance of social structure. And it has important symbolic value as well. In addressing the forces of change acting on the region, and in the context of the changing nature of the relationships between Arabs and Israelis, these distributional issues—which are basically principles of equity—are the key. Adapting to the forces of change noted above will only be possible if equity is integrated within the economic, political, and social context of this change. Indeed, any lasting peace in the region—and any future allocation of water supplies—must be based on equity principles.

REFERENCES

Baumol, W., and W. Oates. 1975. *The Theory of Environmental Policy.* Englewood Cliffs, N.J.: Prentice Hall.

Boulding, K. 1964. "The Economist and the Engineer." In S. Smith and E. Castle (eds.), *Economics and Public Policy in Water Resources Development.* Ames: Iowa State University Press.

Gleick, P. 1990. "Environment, Resources and International Security and Politics." In E. H. Arnett (ed.), *Science and International Security: Responding to a Changing World,* pp. 501–523. Washington, D.C.: American Association for the Advancement of Science.

Hillel, D. 1995. *Rivers of Eden.* Oxford: Oxford University Press.

Isaac, J., and A. Kubursi. 1996. *Dry Peace in the Middle East?* Bethlehem: Applied Research Institute of Jerusalem.

Israeli Bureau of Statistics. 1994. *Statistical Abstract of Israel.* Jerusalem: Central Bureau of Statistics.

Kliot, N. 1994. *Water Resources and Conflict in the Middle East.* London: Routledge.

Lonergan, S. C., and D. Brooks. 1994. *Watershed: The Role of Freshwater in the Israeli-Palestinian Conflict.* Ottawa: IDRC Press.

Lonergan, S. C., and B. Kavanagh. 1991. "Climate Change, Water Resources and Security in the Middle East." *Global Environmental Change* 1 (4): 272–290.

Rogers, P. 1994. "The Agenda for the Next Thirty Years." In P. Rogers and P. Lydon (eds.), *Water in the Arab World,* pp. 285–316. Cambridge, Mass.: Harvard University Press.

Roots, E. F. 1992. "International Agreements to Prohibit or Control Modification of the Environment for Military Purposes: An Historical Overview." In H. B. Schiefer (ed.), *Verifying Obligations Respecting Arms Control and the Environment: A Post Gulf War Assessment.* Workshop Proceedings, Arms Control and Disarmament Division, External Affairs and International Trade Canada, Ottawa.

United Nations. 1994. *World Population Prospects.* New York: United Nations.

Wishart, D. 1993. "Shock Therapy for Middle East Water Managers." In *Proceedings of the International Symposium on Water Resources in the Middle East: Policy and Institutional Aspects,* pp. 69–75. Urbana, Illinois, October 1993.

Wolf, A. 1995. *Hydropolitics along the Jordan River: Scarce Water and Its Impact on the Arab-Israeli Conflict.* Tokyo: United Nations University Press.

World Bank. 1993. *Developing the Occupied Territories: An Investment in Peace.* Washington, D.C.: World Bank.

———. 1996. *World Development Report, 1996.* Washington, D.C.: World Bank.

World Resources Institute. 1996. *World Resources, 1996–1997.* New York: Oxford University Press.

4. *"Hydrostrategic" Territory in the Jordan Basin*
WATER, WAR, AND ARAB-ISRAELI
PEACE NEGOTIATIONS

Aaron T. Wolf

INTRODUCTION

This chapter examines the relationship between the location of water sources and strategic territory along Arab-Israeli boundaries, and poses the question, "Does territory exist over which sovereignty has been sought politically or militarily, or which would be insisted upon in the course of current territorial negotiations, *solely* because of its access to water sources, and in the absence of any other compelling strategic or legal rationale?" The question as a whole is divided into three components:

1. Have boundaries been drawn historically on the basis of the location of water access? It is found that water sources have played a role, albeit subservient to other concerns, in the delineation of international boundaries, first between the British and French Mandates, then among Israel, Lebanon, and Syria. In particular, the political and military policymakers of Israel had explicit interests in retaining the northern headwaters of the Jordan River, arguing for them in political arenas and reinforcing claims through settlement policy. Yet it is also clear that once boundaries were agreed to in a legal forum in 1923, development plans were modified to fit the legal boundaries, *not* vice versa.

2. During warfare between competing riparians, has territory been explicitly targeted or captured because of its access to water sources? Despite a growing literature which suggests that Israeli-Arab warfare has had a "hydrostrategic" component, the evidence suggests that water resources were not at all factors for strategic planning in the hostilities of 1948, 1967, 1978, or 1982. By this I mean that the decision to go to war, and strategic decisions made during the fighting, including which territory was necessary to capture, were not influenced by water scarcity or the location of water resources.

3. In the course of negotiations, has territory with access to water sources, and no other strategic component, been seen as vital to retain by any of the riparians? The questions of water allocations and rights have been difficult components in the Arab-Israeli peace talks. Nevertheless, with the concluded negotiations between Israel and Jordan, and the ongoing talks between Israel and the Palestinians, and despite the quantity of studies identifying hydrostrategic territory and advising its retention, *no* territory to date has been retained simply because of the location of water. Solutions in each case have focused on creative joint management of the resource, rather than insistence on sovereignty.

"Water" and "war" are two topics being assessed together with increasing frequency. Articles in the academic literature (Cooley 1984; Starr 1991; Gleick 1993; and others) and popular press (Bulloch and Darwish 1993; *World Press Review* 1995) point to water not only as a cause of historic armed conflict, but as *the* resource which will bring combatants to the battlefield in the twenty-first century. Invariably, these writings on "water wars" point to the arid and hostile Middle East as an example of a worst-case scenario, where armies have in fact been mobilized and shots fired over this scarce and precious resource. Elaborate, if misnamed, "hydraulic imperative"[1] theories have been developed for the region, particularly regarding conflict between Arabs and Israelis, citing water as the prime motivator for military strategy and territorial conquest.

The basic argument is as follows: Water is a resource vital to all aspects of a nation's survival, from its inhabitants' biology to their economy; the scarcity of water in an arid environment leads to intense political pressures, often referred to as "water stress" (a term coined by Falkenmark 1989); the Middle East is a region not only of extreme political conflict, but also of states that are reaching the limits of their freshwater supply; therefore, Middle East warfare and territorial acquisition *must* be related to the region's "water stress" (see Figure 4.1).

The detailed assessment of the validity of a "hydraulic imperative" leading to warfare has importance beyond the regional setting of the Jordan basin. Those in the growing study of "environmental security" who would claim future "water wars" around the globe invariably use conflict along the Jordan as a case study where water *has* led to war—Myers (1993), for example, devotes an entire chapter to water in the Middle East as his first example of "ultimate security"; and Homer-Dixon (1994), citing the Jordan and other water disputes, comes to the conclu-

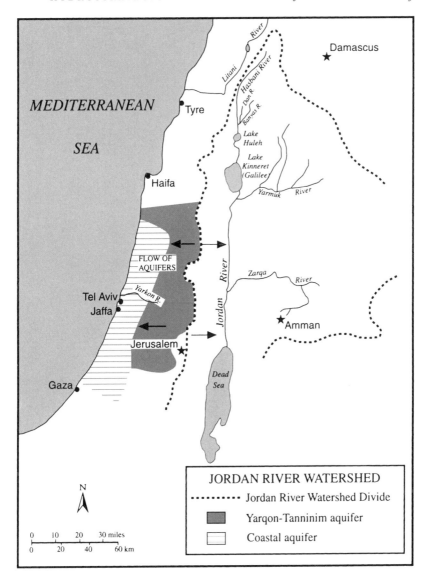

FIGURE 4.1. Jordan River Watershed (Redrawn from Wolf 1995a).

sion that "the renewable resource most likely to stimulate interstate resource war is river water."

More importantly, views of historic conflict inevitably inform the parameters of conflict-resolving negotiations. Water has been a vital component of ongoing negotiations between Arabs and Israelis, as described

by Hof (1995) and Wolf (1995b). Each stage of negotiations has been accompanied by a chorus of editorial and academic advice urging the retention of territory seen as essential to the hydrologic survival of one side or another. This will no doubt be true for future negotiations as well. It is vital to these future rounds that a clear exploration of water's actual relationship to territory be undertaken.

This chapter, then, seeks to examine in detail the premise of a link between water and land—the nature of "hydrostrategic territory"—in this "worst-case" water conflict between Arabs and Israelis.

In order to answer the questions posed above, we must first define our terms, then distinguish "hydrostrategic" territory—that is, land surface which has strategic value solely for its access to water resources—from other territory under conflict. The linking of international conflict to the attributes of border regions has been well documented, and water, as a unique natural resource, has caused its disproportionate share of boundary conflicts. Prescott (1965, 126–133) describes disputes over water bodies which mark or cross a boundary (including territorial waters) as the most common source of functional disputes, and both he and Bingham, Wolf, and Wohlgenant (1994) describe such examples from around the world. Cohen (1986) lists almost one hundred boundary disputes, and describes the most important elements which contribute to them, including some most relevant to this study: strategic and tactical land space; strategic water space; land access to the sea; strategic minerals; water resources for irrigation, drinking, and electric power; historic claims; drives for racial, ethnic, or religious unification; minority struggles for independence; refugee populations; distractions from domestic turmoil; and major power rivalries.

It can be difficult to distinguish between military strategy, defined concisely by one military officer as "From where are they shooting and from where will we shoot back?" (cited in Wolf 1995a, 73), and "hydrostrategy," the influence of the location of water resources on strategic thinking. A river, for example, is also a valuable barrier against tanks and troop movements, and, as clear landmarks, rivers often delineate boundaries. High ridges, ideal for military positioning, are also often local watershed boundaries. As Minghi (1963) points out, the most thought was given to the relationship between boundaries and security, particularly in the periods including the two world wars—times of intense boundary delineation.

It is precisely these conflicting elements of water sources—their strategic value in the traditional sense, their functional value in a domestic sense, and their role in delineating boundaries—which inform the central questions of this chapter posed above. For the purposes of this study, then, I define "hydrostrategic territory" as that territory which has strategic value *primarily* because of its access to water resources for irrigation, drinking, and/or electric power. This is distinguished from strategic territory in a traditional military or political sense, including what Cohen (1986) calls "strategic water space," or water-related territory which provides traditional strategic value.

One note: Wolf (1996) distinguishes between a water "crisis," the lack of water in the basin for anticipated needs, and a water "conflict," the political tensions attendant to the lack of water. This chapter is not about the regional water crisis. There are literally thousands of books and articles about the lack of fresh water in the Middle East, and what might be done to alleviate that shortage.[2] My conclusions about territory are "only" about territory, and should in no way be taken to belittle the seriousness of the lack of water in the region.

WATER AND BOUNDARIES

In this section, I seek to answer the first question posed in the introductory section, that is: Have boundaries been drawn historically on the basis of the location of water access?

Boundary Proposals and Delineation, 1913–1923

In this century, as the developing modern nationalisms of both Arabs and Jews became clearly defined, and with subsequent population pressures accelerated by immigration, the boundaries of today's Middle Eastern nations began to take shape. After the first Zionist Congress in Basel in 1897, the idea of creating a Jewish state in Palestine, which by then had been under Ottoman rule for four hundred years, began to crystallize in the plans of European Jewry, and efforts were made, without much result, to gain the support of Turkish or British authorities. Even without commitments for independent nations, both Jewish and Arab populations began to swell in turn-of-the-century Palestine, the former in waves of immigration from Yemen as well as from Europe, and the latter attracted to new regional prosperity from other parts of the Arab world (Sachar 1969; McCarthy 1990). According to McCarthy (1990), Palestine had 340,000 people in 1878 and 722,000 by 1915.

During World War I, as it became clear that the Ottoman Empire was crumbling, the heirs-apparent began to jockey for positions of favor with the inhabitants of the region. The French had inroads with the Maronite Catholics of Lebanon and therefore focused on the northern territories of Lebanon and Syria. The British, meanwhile, began to seek coalition with the Arabs from Palestine and Arabia—whose military assistance against the Turks they desired—and with the Jews of Palestine, both for military assistance and for the political support of Diaspora Jewry. As the course of the war became clear, French and British, Arabs and Jews, all began to refine their territorial interests.

A detailed description of the lengthy process which finally led to the final determination of boundaries for the French and British Mandates, which, in turn, informed the boundaries of modern Lebanon, Syria, Jordan, and Israel, is beyond the scope of this chapter, but can be found in the works of Ra'anan (1955), Sachar (1969, 1987a), and Fromkin (1989). The influence of water resources along the Palestine-Syria border is described by Garfinkle (1994), and along the Palestine-Lebanon frontier by Hof (1985) and Amery (1998). The relationship between boundary demarcation and water development is summarized by Brawer (1968) and Wolf (1995a). The interested reader is referred to that literature for more detail, and the lines which resulted are shown in Figure 4.2.

In summary, water only began to influence boundaries with the Paris Peace Talks in 1919. The Zionists began to formulate their desired boundaries for the "national home," to be determined by three criteria: historic, strategic, and economic considerations (Zionist publications cited in Ra'anan 1955, 86). Historic concerns coincided roughly with British allusions to the biblical "Dan to Beersheba." These were considered minimum requirements which had to be supplemented with territory that would allow military and economic security. Military security required desert areas to the south and east as well as the Beka'a Valley, a gateway in the north between the Lebanon Mountains and Mount Hermon.

Economic security was defined by water resources. The entire Zionist program of immigration and settlement required water for large-scale irrigation and, in a land with no fossil fuels, for hydropower. The plans were "completely dependent" on the acquisition of the "headwaters of the Jordan, the Litani River, the snows of Hermon, the Yarmuk and its tributaries, and the Jabbok" (Ra'anan 1955, 87).

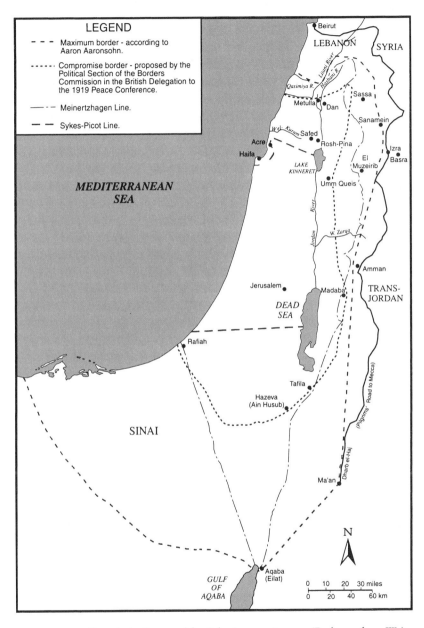

LEGEND

FIGURE 4.2. Boundaries Proposed for Palestine, 1916–1919 (Redrawn from Weizmann 1968).

Aaron Aaronson's "The Boundaries of Palestine" (27 January 1919, unpublished, Zionist Archives), drafted in less than a day, argued that:

In Palestine, like in any other country of arid and semi-arid character, animal and plant life and, therefore, the whole economic life directly depends on the available water supply. It is, therefore, of vital importance not only to secure all water resources already feeding the country, but also to insure the possession of whatever can conserve and increase these water—and eventually power—resources. The main resources of Palestine come from the North, from the two mighty mountain-masses—the Lebanon range, and the Hermon. . . .

The boundary of Palestine in the North and in the North East is thus dictated by the extension of the Hermon range and its water basins. The only scientific[ally] and economic[ally] correct lines of delineation are the water-sheds.

Aaronson then described the proposed boundaries in detail, as delineated by the local watershed (see Figure 4.2). In June 1919, Aaronson died in a plane crash on his way to the Peace Conference. Nevertheless, the importance of the region's water resources remained embedded in the thinking of the Zionist establishment. "So far as the northern boundary is concerned," wrote Chaim Weizmann later that year, "the guiding consideration with us has been economic, and 'economic' in this connection means 'water supply'" (Weizmann Letters [Weizmann 1968], 18 September 1919).

The Arab delegation to the Peace Conference was led by the Emir Feisal, younger son of Emir Hussein of the Hejaz. Working with T. E. Lawrence, Hussein and his sons had led Arab irregulars against the Turks in Arabia and Eastern Palestine. The Arab requests were spelled out in a memorandum dated 1 January 1919. Because the territory in question was so large—including Syria, Mesopotamia, and the Arabian Peninsula—geographically diverse, and, for the most part, well watered, it is not surprising that water resources played little role in the Arab deliberations. Based on a combination of level of development and ethnic considerations, Feisal (from Esco Foundation 1947) asked that: Syria, agriculturally and industrially advanced, and considered politically developed, be allowed to manage its own affairs; Mesopotamia, "underdeveloped and thinly inhabited by semi-nomadic peoples, [be] buttressed . . . by a great foreign power," but governed by Arabs chosen by the "selective rather than the elective principle"; the Hejaz and Arabian

Peninsula, mainly a tribal area suited to patriarchal conditions, retain their complete independence.

Two areas were specifically excluded: Lebanon—"because the majority of the inhabitants were Christian"—which had its own delegates (see Amery 1998 for details regarding Lebanon), and Palestine, which, because of its "universal character was left to one side for mutual consideration of all parties interested" (Esco Foundation 1947, 138).

Once testimony was heard at Versailles, the decisions were left to the British and the French, as the peace talks continued, culminating at San Remo in 1920, as to where the boundaries between their Mandates would be drawn. In late 1919, the British first suggested the "Meinertzhagen line" as a boundary. This line, by following the right bank rather than the thalweg, included much of the river system (see Figure 4.2.). The Meinertzhagen line was similar in the north to the Zionist proposals, and was rejected by the French.

In September the British put forward the compromise "Deauville Proposal," which granted Palestine less territory than the Zionists sought, but still included the southern bank of the Litani and the Banias headwaters. Finally, to meet French objections as far as possible, the British proposed a border running north from Acre to the Litani bend, then east to Mount Hermon, which would increase Lebanese territory but leave the headwaters in Palestine.

Although the French rejected each of these proposals, Phillips Berthelot, the foreign minister and negotiator at an Anglo-French conference on the Mideast in December 1919, suggested that Prime Minister Georges Clemenceau insisted on the Sykes-Picot line, but that he was prepared

> to agree that one-third of the waterpower of the waters flowing from Mount Hermon southwards into the Palestine of the Sykes-Picot agreement should be allotted to the Zionists under an economic arrangement with France. The French could do no more than this. (Ra'anan 1955, 125)

In June 1920, France agreed to a compromise: Palestine's northern boundary should be a line drawn from Ras en-Naqura to a point on the Jordan just north of Metulla and Banias-Dan, and then to the northern shore of Huleh Lake, running from there along the Jordan, down the middle of the Sea of Galilee to the Yarmuk, where it would meet the Sykes-Picot line. Although these boundaries included all existing Jewish settlements within Palestine, most of the water resources would remain in Syria.

At the San Remo Conference in April 1920, agreement was reached where Great Britain was granted the Mandates to Palestine and Mesopotamia, and France received the Mandate for Syria (including Lebanon). During the remainder of the year, last-minute appeals were made both by the British and by the Zionists for the inclusion of the Litani in Palestine or, at the least, for the right to divert a portion of the river into the Jordan basin for hydropower. The French refused, offering a bleak picture of the future without an agreement, and suggested, referring to British and Zionist ambiguity as to what was meant by a "National Home," "Vous barbotterez si vous le voulez, mais vous ne barbotterez pas à nos frais"[3] (Butler and Bury 1958, 8:387).

On 4 December 1920, a final agreement was reached in principle on the boundary issue, which addressed, mainly, French and British rights to railways and oil pipelines, and incorporated the French proposal for the northern boundaries of six months previously. The French delegation did promise that the Jewish settlements would have free use of the waters of the Upper Jordan and the Yarmuk, although they would remain in French hands. The Litani was excluded from this arrangement. Article 8 of the Franco-British Convention, therefore, included a call for a joint committee to examine the irrigation and hydroelectric potential of the Upper Jordan and Yarmuk, "after the needs of the territories under French Mandate," and added that:

> In connection with this examination the French government will give its representatives the most liberal instructions for the employment of the surplus of these waters for the benefit of Palestine. (Hof 1985, 14)

The final boundaries between the French and British Mandates, which later became the basis for the boundaries between Israel, Lebanon, Syria, and Jordan, were worked out by an Anglo-French commission set up to trace the frontier on the spot. Biger (1989) cites the principles followed by the commission as generally circumventing settlements, dividing ethnically (based on the wishes of the inhabitants), and retaining the link between settlements and their agricultural land. Its conclusions were submitted in February 1922 and signed by the British and French governments in March 1923 (see Figure 4.3).

The frontier would run from Ras en-Naqura inland in an easterly direction along the watershed between the rivers flowing into the Jordan and into the Litani; the line was then to turn sharply north to include in

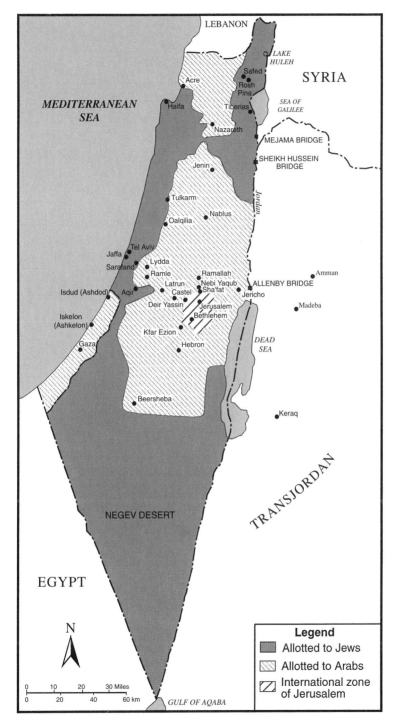

FIGURE 4.3. United Nations Plan for the Partition of Palestine, November 1947 (Redrawn from Sachar 1979).

Palestine a "finger" of territory near Metulla and the eastern sources of the Jordan.

Rather than include the Banias Spring within Palestine, as in the French proposal of six months previously, the border ran parallel to and 100 meters south of the existing path from Metulla to the Banias. The French insisted on inclusion of this road in its entirety to facilitate east-west transportation and communication within its Mandate—this is a stretch of the main route from Tyre to Quneitra (Brawer 1968). This northern border meant that the entire Litani and the Jordan headwaters of the Ayoun and Hasbani would originate in Lebanon before flowing into Palestine. The Banias Spring, meanwhile, would originate and flow for 100 meters in Syrian territory, then into Palestine. Since Palestine had a promise of water use, and also access to the Banias Heights, which overlooked the spring, the fact that the actual spring lay outside of the boundaries was not of immediate concern. Of the headwaters of the Jordan, however, only the Dan Spring remained entirely within Palestine.

From Banias, the border turned south toward the Sea of Galilee, along the foothills of the Golan Heights, parallel and just east (sometimes within 50 meters) of the Huleh Lake and the Jordan River. Rather than passing through the middle of the Sea of Galilee, the border ran 10 m east of its shores (even if the level should rise because of a proposed dam), leaving the entire lake, the town of El-Hama, and a small triangle just south of the Jordan's outflow within the territory of Palestine. These latter two were already included in Zionist plans for water diversion and hydroelectricity generation. These changes were beneficial to Palestine's hydrostrategic positioning, although they were made mainly for administrative reasons, "to make customs inspection easier." Nevertheless, according to the agreement, fishing and navigation rights on the lake were retained by the inhabitants of Syria. Later, between 1948 and 1967, Syria claimed the shoreline as the de facto border, and argued for riparian rights to the lake on that basis.

At the Yarmuk, the border went eastward along the river, meeting up with the Sykes-Picot line into the Syrian desert and south of the Jebel Druze.

The final agreement made no mention of joint access to French-controlled waters.

Administrative divisions would further convolute the boundaries along the Jordan. In May 1921, Winston Churchill offered the Emir Abdullah reign over that part of the British Mandate east of the Jordan River—Transjordan became a semi-independent entity in 1923, and an independent kingdom in 1946. While the division was originally be-tween administrative units under one authority—the British Mandate—the boundary, as defined by the center of the Jordan River, the Dead Sea, and Wadi Araba/Arava, would eventually become the international boundary between Israel and Jordan.[4]

Although the location of water resources had been an important, sometimes overriding issue with some of the actors involved in determin-ing the boundaries of these territories, it is clear in the outcome that other issues took precedence over the need for unified water basin de-velopment. These other factors ranged from the geostrategic—the loca-tion of roads and oil pipelines—to political alliances and relationships among British, French, Jews, and Arabs, to how well-versed one or an-other negotiator was in biblical geography.

Partition and Statehood: 1922–1948

During the 1930s and 1940s, water was a focus of several reports which tried to determine the "economic absorptive capacity" of the land. In the absence of clear immigration policy, both Jewish and Arab residents of Palestine became increasingly frustrated, taking out their hostility on each other as well as on the British. Over time, the partition of Palestine into Jewish and Arab states increasingly became the most advocated op-tion, first in an Anglo-American plan in 1946, and later, when Britain ceded the Mandate to the United Nations, in the UN Partition Plan of 1947 (see Figure 4.3).

In the case of partition, it became clear to the Zionists that, at a mini-mum, three areas were needed for a viable Jewish state: the Galilee re-gion with the Jordan headwaters, the coastal zone with the population centers, and the Negev Desert, to absorb "the ingathering of the exiles." In the late 1930s, the Jewish Agency, sensing that partition was immi-nent, set out on an intensive settlement program building fifty-five farm communities between 1936 and 1939 (Sachar 1979, 216). The emphasis for site location was in the northern Galilee, to reinforce the projected boundaries and to guarantee the inclusion of what Jordan headwaters were left from the Mandate process.

The Zionist position on whether partition should occur and, if so, what the minimum territorial requirements would be for a viable Jewish state was increasingly influenced by Walter Clay Lowdermilk. Lowder-milk, director of the U.S. Soil Conservation Service, published in 1944 *Palestine: Land of Promise* at the commission of the Jewish Agency. In contrast to the Ionides Plan of 1939, Lowdermilk asserted that proper water management would generate resources for 4 million Jewish refugees in addition to the 1.8 million Arabs and Jews living in Palestine at the time. He advocated regional water management, based on the Tennessee Valley Authority (TVA), to develop irrigation on both banks of the Jordan River and in the Negev Desert, and building a canal from the Mediterranean to the Dead Sea to generate hydropower and replenish the diverted fresh water (Naff and Matson 1984, 32).

At the same time, a 1944 study, *The Water Resources of Palestine*, undertaken by Mekorot, the national water company for Jewish Palestine, described an "All-Palestine Project" for irrigation and hydroelectric development. The study included frontier adjustments which would be desirable for a basinwide development scheme in Palestine (see Figure 4.4). It was suggested that the Mandate border be moved upstream where it would meet the Hasbani, Dan, and Banias headwaters to allow for more effective drainage; eastward along Huleh Lake to leave room for a conduit on the east side of the lake; and upstream along the Yarmuk to include an area of about 80 km² of Transjordan to develop a series of impoundments along the river (Mekorot 1944). It should be noted that, although the report included plans to bring Litani water into the Jordan watershed, it was assumed that agreement would have to be reached with the Lebanese government to do so. Lebanese territory was not included in the list of desirable frontier adjustments. It should also be noted that there is no evidence that any of the territorial suggestions of the Mekorot study were ever included in political decision-making, nor were the proposed boundary modifications raised in subsequent negotiations.

Water and Boundaries—Conclusions

Prescott (1987, 94) notes that the most common cause of boundary disputes can be found in the history of the boundary—that the "evolution" of the boundary is "incomplete." In answer to the question of water's influence on boundaries posed at the beginning of this section, it is clear that water sources have played a role, albeit subservient to other concerns, in the delineation of international boundaries, first between the British and French Mandates, then among Israel, Lebanon, and Syria. In particular, the political and military policymakers of Israel had explicit interests in retaining the northern headwaters of the Jordan River, arguing for them in political arenas and reinforcing claims through

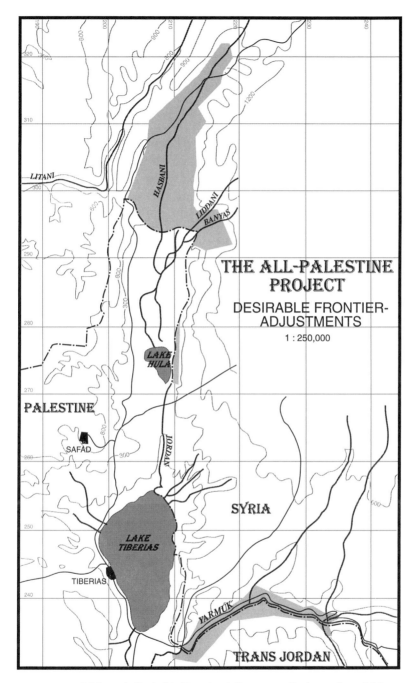

FIGURE 4.4. Mekorot's Desirable Frontier Adjustments (Redrawn from Mekorot 1944).

settlement policy. The goal, however, was reached with only marginal success. While the headwaters of the Jordan originated in the territories of three separate entities, Israel had been successful in retaining riparian access to both banks of the Jordan River above the Sea of Galilee, and sovereignty over the entire lake. Yet it is also clear that once boundaries were agreed to in a legal forum in 1923, development plans were modified to fit the legal boundaries, *not* vice versa.

WATER AND WAR

The next aspect of "hydrostrategic territory" to be addressed is the relationship between military strategic thinking during wartime and the location of water resources. In this section, I try to answer the question, "During warfare between competing riparians, has territory been explicitly targeted, captured, or retained because of its access to water sources?" This has been the most sensitive and, I will argue, the most misinterpreted of the three components of this chapter. Because of possible repercussions for interpretation of the region's history, for strategies for future negotiations, and for extrapolation into the broader world of environmental security, I examine the link between water and war in great detail.

Background: Evidence of a "Hydrostrategic Imperative"?

In recent years, particularly since Israel's invasion of Lebanon in 1982, a "hydraulic imperative" theory, which describes the quest for water resources as *the* motivator for Israeli military conquests, both in Lebanon in 1979 and 1982 and earlier, on the Golan Heights and West Bank in 1967, was developed in the academic literature and the popular press. Proponents of the theory, which might be better termed the "hydrostrategic imperative,"[5] usually point to some combination of the following facts to support their argument (see, for example, Davis et al. 1980; Stauffer 1982; Schmida 1983; Stork 1983; Cooley 1984; Dillman 1989; Beaumont 1991):

· Early Zionist lobbyists and planners, from Chaim Weizmann at the Paris Peace Conference in 1919 through Hays and Cotton and their plans of the 1940s and 1950s, have advocated inclusion either of the Litani River within Israeli boundaries or of Litani water in the Jordan River watershed.

- The 1967 war had been precipitated by tensions over Israeli and Arab water diversion schemes, and the war itself had greatly strengthened Israeli "hydrostrategic" positioning.
- The 1979 Litani Operation left Israel and its allies in control of the lower Litani River and the Hasbani headwaters.
- The 1982 Israeli invasion of Lebanon included capture of the Qir'awn Dam and related data. Even after Israel's redeployment in 1985, the "Security Zone" still leaves Israel in control of the area from Taibe and slightly north—the most likely site for any Litani-to-Jordan-basin transfer.

Particularly during the years of Israeli occupation from 1982 to 1985, several analysts extrapolated from this history to speculate on likely Israeli actions in Lebanon. Proponents of this theory proposed scenarios ranging from a simple diversion of the 100 mcm/yr available at the lower Litani, to elaborate conjectures of a permanent occupation of the entire Beka'a Valley south of the Beirut-Damascus Highway which, along with a hypothetical destruction of the Qir'awn Dam and Marhaba diversion tunnel and forced depopulation of southern Lebanon, would allow diversion of the entire 700-mcm/yr flow of the river into Israel.[6]

Others have argued that Israel retains access to the Litani through its "Security Zone" because it is, in fact, covertly diverting water into the Jordan basin. According to John Cooley, "It was small wonder that the first Israeli diversion plans for the Litani have come into being" (cited in Soffer 1994, 6). More recently, Beaumont (1994, 15) claims that Israel "may well be stealing Lebanese water for its own use." Frey and Naff (1985, 76), even while arguing against the imperative, do suggest that:

> Although water may not have been the prime impetus behind the Israeli acquisition of territory . . . it seems perhaps the main factor determining its retention of that territory.

Prof. Thomas Naff later testified to Congress that "Israel is presently conducting a large-scale operation of trucking water to Israel from the Litani River" (U.S. House of Representatives 1990, 24). Naff (1992, 6) has since modified the contention to, "Water, it seems, was instead trucked to units of the Israeli-supported Lebanese Army of South Lebanon in the same area as a reward for their cooperation."

Beaumont (1991, 8) is typical of those who, building retroactively on the charges regarding Lebanon, now include the 1967 war as proof of water driving Israel's territorial "imperative":

To avoid each of the states [Lebanon and Syria] controlling their own water resources, Israel invaded southern Lebanon[7] and the Golan Heights of Syria in 1967. The pretext given was strategic reasons, but the control of the water resources of the area seems a more compelling and realistic reason.

The theory that water has driven strategic thinking during wartime has been critiqued for political and technical weaknesses by Naff and Matson (1984, 75–80), Wolf (1995a, 70–80), Soffer (1994), and Libis-zewski (1995), as well as on economic grounds by Wishart (1989, 14), yet, as evidenced by the passages above, seems still to survive. To examine the validity of a "hydrostrategic" imperative, two questions must be answered: Was the location of water resources a factor in the military strategy of Israel in 1967, 1978, or 1982? Is Israel now diverting water from the Litani River?

The Arab-Israeli Wars: 1948, 1967, and 1982

The War of 1948

On 2 February 1947, Great Britain officially turned the fate of Palestine over to the United Nations. The UN Special Committee on Palestine recommended partition of Palestine into two states, but included a vehicle for joint economic development, "especially in respect of irrigation, land reclamation, and soil conservation" (see Figure 4.3).

The Jewish state included the areas described above, and the Arab state included the remainder of Palestine, based on population centers. Jerusalem was to be an international city, and the Jewish state would pay a £4 million annual stipend to the Arab state to reflect the more advanced agricultural and industrial position of the former (UN Resolution on the Partition of Palestine 1947, Chapter 4). The General Assembly approved the Partition Plan on 29 November 1947.

Though the Jewish Agency reluctantly accepted partition, the Arab states rejected it outright and, when the British pulled out of Palestine in May 1948, Egypt, Jordan, Iraq, Syria, Lebanon, and Saudi Arabia went to war against the new state of Israel. During the 1948 war, keeping the three zones described above as necessary for a viable Jewish state—the Galilee region with the Jordan headwaters, the coastal zone with the population centers, and the Negev Desert, to absorb anticipated immigration—became the focus for the Israeli war effort.

The Israelis lost three other strategic points along waterways, though,

FIGURE 4.5. Rhodes Armistice Demarcation Line, 1948 (Redrawn from Sachar 1979).

and the repercussions would be felt through 1967 (see Figure 4.5 and Figure 4.6). As mentioned above, during the Mandate negotiations, the French had denied the Zionists the Banias Spring because an access road they needed crossed the waterway about 100 meters downstream. To guarantee access to the water, though, a small hill overlooking the stream, Givat Banias, had been included in Palestine. Second, the town of El-Hama, located on one of the few flat areas within the narrow Yarmuk Valley, and an adjacent triangular area from the Yarmuk to the

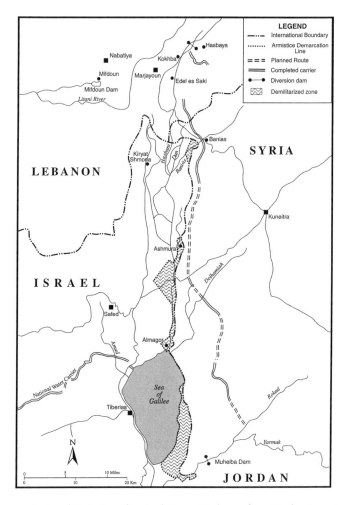

FIGURE 4.6. Israel-Syria Demilitarized Zones (Redrawn from Medzini 1976).

shores of the Sea of Galilee were included within Palestine's territory, specifically to facilitate water resources development. Both of these areas were lost to the Syrians during fighting in 1948 (Sachar 1979). Also, the Syrians crossed the Jordan River and occupied land in the Huleh Lake– Daughters of Jacob Bridge region, an area targeted by water planners as the site of the first major Israeli reclamation project. Finally, although the Israeli army had occupied a strip of Lebanese territory along the elbow of the Litani, it pulled back to the Mandate boundaries as part of the armistice agreement, in the unfulfilled hope of gaining a peace treaty with Lebanon (Hof 1985, 31).

The 1948 war added a new type of boundary to the region—the Armistice Line. While the boundary between the British and French Mandates had the permanence and force of international law, the armistice agreement, signed separately in Rhodes between Israel and each of its neighbors between February and July 1949, was explicitly stated *not* to constitute political boundaries—only to delineate a temporary military agreement (Hareven 1977). Negotiations continued in an unsuccessful quest for a permanent peace agreement, culminating in Lausanne late in 1949 (Caplan 1993). Water was not mentioned at all during the Rhodes Armistice Talks, and only intermittently during the Lausanne Conference.[8]

The differences between permanent legal boundaries and the Armistice Line were manifested in how each boundary segment was treated by each combatant:

Israel-Lebanon Boundary

Israel had occupied Lebanese territory up to the Litani River, yet withdrew its forces to the international boundary as a result of the armistice agreement. Amery (1998, 23) suggests that Israel's withdrawal was based on the belief that it could make peace with a Christian-led Lebanon and that joint Lebanese-Israeli water resources development could proceed without territorial annexation. Amery (1998, 23) cites Berger (1965, 30) as arguing that Israel would not have withdrawn from southern Lebanon had it not been convinced of these results.

Israel-Syria Boundary

Syria occupied about 60 km[2] of Israeli territory during the war at three locations, as described above: the Banias Springs area, the Huleh Lake–Daughters of Jacob Bridge region, and the triangle from El-Hama to the southeastern shores of the Sea of Galilee. During the armistice talks, Syria agreed to withdraw from all of this territory except the Givat Banias hill and the town of El-Hama, provided the remaining territory not be militarized by Israel. While both of these sites which Syria retained were included in Israeli water development plans, Israel did not push for the return of this relatively small territory which deviated from the international boundary, given the presumed temporary nature of the Armistice Line.

Neff (1994, 27) suggests that Syria withdrew with the understanding that final borders, including final sovereignty over the three demilitarized zones (DMZs), would be negotiated in the future. Israel, in contrast, considered itself the legal sovereign of these areas, legal heir to the northern territory within the British Mandate.

Israel-Jordan Boundary

The prewar boundary between British-Mandate Palestine and Transjordan, delineated when Britain split Palestine in two in 1922, followed the middle of the Yarmuk River, the Jordan River, the Dead Sea, and Wadi Araba/Arava southward to the Gulf of Aqaba/Eilat (Biger 1994). After the war, Jordan claimed jurisdiction over the West Bank, and nowhere is the assumption that the Armistice Line was to be temporary more clear than in the "Green Line," the boundary between the West Bank and Israel. While negotiations continued officially in Rhodes, Sachar (1979, 349–350) describes secret meetings which took place directly between the Israelis and King Abdullah and his advisors at the king's winter palace. The agreement which was reached was informed not only by the location of the two armies at war's end, but also by the location of roads and railways; and by hilltops and high ground for local strategic advantage, based on the type of weaponry available to each side. The line, drawn in green on a map at a scale of 1 : 250,000,[9] cut villages from their land, divided towns from the springs on which they relied, and occasionally split settlements in two (Biger 1989). As Biger (1989) points out, "in no case did the terms of agreement provide for continuing rights of access by inhabitants to their vital land and water resources."

As a result of the 1948 war, the Jordan River was even more divided than it had been under the Mandates. The Hasbani rose in Lebanon with the Wazzani, a major spring of the Hasbani, situated only a few kilometers north of the Israeli border. The Banias flowed for 100 meters in Syrian territory before crossing into Israel. The Dan rose and remained within Israeli territory. The confluence of the three, the Jordan River, flowed along the Israeli-Syrian border, often through a demilitarized zone, until it reached the Sea of Galilee. The sea lay wholly in Israel, with the Syrian border 10 meters from the eastern coast. The Yarmuk rose in Syria, then became the Syrian-Jordanian border until its confluence with the Jordan. South of the Sea of Galilee, the Jordan River formed first the Israeli-Syrian border, then the Israeli-Jordanian border below the confluence with the Yarmuk, finally flowing wholly into Jordanian territory and the Dead Sea, which was about one-quarter Israeli and three-quarters Jordanian. Groundwater was equally divided, with the recharge zones of two springs on which Israel increasingly relied, the Yarkon and the Taninim, originating in the Jordanian territory of the West Bank (see Figure 4.5).

The War of 1967

Water Resources and Background to the War

For the Jordan River, the legacy of the Mandates, and the 1948 war, was a river divided in a manner so convoluted that unilateral water resources development, the only strategy available to these hostile riparians, would lead inevitably to conflict. By the early 1950s, Arab states were discussing organized exploitation of two northern sources of the Jordan—the Hasbani and the Banias (Stevens 1965, 38). The Israelis also made public their All Israel Plan, which included the draining of Huleh Lake and swamps, diversion of the northern Jordan River, and construction of a carrier to the coastal plain and Negev Desert—the first out-of-basin transfer for the watershed (Naff and Matson 1984, 35).

Jordan, in 1951, announced a plan to irrigate the East Ghor of the Jordan Valley by tapping the Yarmuk. At Jordan's announcement, Israel closed the gates of an existing dam south of the Sea of Galilee and began draining the Huleh swamps, which lay within the demilitarized zone with Syria. These actions led to a series of border skirmishes between Israel and Syria which escalated over the summer of 1951 (Stevens 1965, 39). In July 1953, Israel began construction on the intake of its National Water Carrier at the Daughters of Jacob Bridge north of the Sea of Galilee and in the demilitarized zone. Syria deployed its armed forces along the border and artillery units opened fire on the construction and engineering sites (Cooley 1984, 3, 10). Syria also protested to the UN and, though a 1954 resolution for the resumption of work by Israel carried a majority, the USSR vetoed the resolution. The Israelis then moved the intake to its current site at Eshed Kinrot on the northwestern shore of the Sea of Galilee (Garbell 1965, 30; see also Figure 4.7).

Against this tense background, President Dwight Eisenhower sent his special envoy Eric Johnston to the Middle East in October 1953 to try to

FIGURE 4.7. International Borders, 1948–1967 (Redrawn from Wolf 1995a).

mediate a comprehensive settlement of the Jordan River system allocations. Johnston's initial proposals were based on a study carried out by
Charles Main and the Tennessee Valley Authority (TVA), at the request
of the UN, to develop the area's water resources and to provide for refugee resettlement (Main 1953). Both Israel and a united Arab League
Technical Committee responded with their own counterproposals. The
Israeli "Cotton" plan included integration of the Litani River's flow into
the Jordan basin, with a subsequent increase in allocations to Israel. The

Arab plan rejected integration of the Litani and substantially reduced Israel's share as compared with the Main plan. Johnston worked until the end of 1955 to reconcile these proposals in a Unified Plan amenable to all of the states involved. In the Unified Plan, Johnston accomplished no small degree of compromise. Though they had not met face-to-face for these negotiations, all states agreed on the need for a regional approach. Israel gave up on integration of the Litani and the Arabs agreed to allow out-of-basin transfers. The Arabs objected, but finally agreed, to storage at both the Maqarin Dam and the Sea of Galilee so long as neither side would have physical control over the share available to the other. Israel objected, but finally agreed, to international supervision of withdrawals and construction. Allocations under the Unified Plan, later known as the Johnston Plan, included 400 mcm/yr to Israel, 720 mcm/yr to Jordan, 35 mcm/yr to Lebanon, and 132 mcm/yr to Syria (Wolf 1995a).

The technical committees from both sides accepted the Unified Plan, but forward momentum died out in the political realm, and the plan was never ratified. Nevertheless, Israel and Jordan have generally adhered to the Johnston allocations, and technical representatives from both countries have met from that time until the present two or three times a year at "Picnic Table Talks," named for the site at the confluence of the Yarmuk and Jordan Rivers where the meetings are held, to discuss flow rates and allocations.

As each state developed its water resources unilaterally, their plans began to overlap. By 1964, for instance, Israel had completed enough of its National Water Carrier that actual diversions from the Jordan River basin to the coastal plain and the Negev were imminent. Although Jordan was also about to begin extracting Yarmuk water for its East Ghor Canal, it was the Israeli diversion which prompted President Gamal Abdel Nasser to call for the First Arab Summit in January 1964, including heads of state from the region and North Africa, specifically to discuss a joint strategy on water.

The options presented at the summit were to complain to the UN, divert the Upper Jordan tributaries into Arab states, as had been discussed by Syria and Jordan since 1953, or to go to war (Schmida 1983, 19). The decision to divert the rivers prevailed at a Second Arab Summit in September 1964, and the Arab states agreed to finance a Headwater Diversion Project in Lebanon and Syria and to help Jordan build a dam on the Yarmuk. They also made tentative military plans to defend the diversion project (Shemesh 1988, 38).

In 1964, Israel began withdrawing 320 mcm/yr of Jordan water for its National Water Carrier, and Jordan completed a major phase of its East Ghor Canal (Inbar and Maos 1984, 21). In November 1964, the Arab states began construction of their Headwater Diversion Plan to prevent the Jordan headwaters from reaching Israel. The plan was to divert the Hasbani into the Litani in Lebanon and the Banias into the Yarmuk, where it would be impounded for Jordan and Syria by a dam at Mukheiba. The plan would divert up to 125 mcm/yr, cut by 35 percent the installed capacity of the Israeli carrier, and increase the salinity in the Sea of Galilee by 60 ppm (Wolf 1995a). In March, May, and August of 1965, Israeli tanks attacked the diversion works in Syria. The final incident, including both Israeli tanks and aircraft on 14 July 1966, stopped Syrian construction, effectively ending water-related tensions between the two states.

Nevertheless, these events set off what has been called "a prolonged chain reaction of border violence that linked directly to the events that led to the [June 1967] war" (Safran, cited in Cooley 1984, 16). Border incidents continued between Israel and Syria, which next triggered air battles in July 1966 and April 1967, and led finally to all-out war in June 1967.

The 1967 War: A "Hydrostrategic" Connection?

In the events leading up to the 1967 war, it has already been noted in some detail how conflict over water resources between Syria and Israel contributed to tensions leading to the fighting, although the hydrologic aspect ended almost a year before the beginning of the war. The war itself started in the south, well away from sensitive water sources, with Egypt expelling the UN forces in the Sinai and blocking Israeli shipping to Eilat. The Sinai Desert was the first front when war broke out on 5 June 1967, with the straits of Sharm-el-Sheikh the primary objective. This alone suggests that issues other than water played a definitive role in strategic decision-making.

The hydrostrategic points over which Israel gained control during the war were on the West Bank, including the recharge zones of several aquifers, some of which Israel had been tapping into since the 1950s; and on the Golan Heights, including the Banias Springs, which Syria had attempted to divert in 1965, and, farther south, El-Hama and an overlook on the proposed site of the Maqarin Dam. The West Bank was controlled by Jordan, and the Golan Heights by Syria. Before the war, and even in its first days, Israel had agreed not to engage in combat with Jordan, as long as Jordan did not attack. Jordan did launch several artillery barrages in the first days of the war, though, which opened up the West Bank as the second front (Sachar 1979).

Finally, despite attacks from Syria, Defense Minister Moshe Dayan was extremely reluctant to launch an attack on the Golan Heights because of the presence of Soviet advisors, and the consequent danger of widening the conflict (Slater 1991). For the first three days of the war, Dayan held off arguments from several of his advisors, including the CO of the northern command, David Elazar, to launch an attack on the Golan Heights. Finally, a delegation from the northern settlements, which had often experienced Syrian sniping and artillery barrages, traveled to Tel Aviv to ask Dayan to take the Heights to guarantee their security. Only then, on June 9, did Israeli forces launch an attack against Syria (Slater 1991, 277).

In the taking of the Golan Heights, the water sources mentioned above were incidental conquests as Israeli forces moved as far east as Quneitra. The only exception is the taking of the town of Ghajar, an Awali village which had no strategic importance in the military sense in that it neither contained combatants nor was situated in a strategic position. It does, however, directly overlook the Wazzani Springs, which contribute 20–25 mcm/yr to the Hasbani's total annual flow of 125 mcm/yr. During dry summer months, the Wazzani is the only flowing source of the Hasbani. Moreover, Ghajar was the site of the projected dams for the Arab diversion project.

It turns out that Ghajar was not even taken during the war. During the fighting, Israeli troops stopped directly outside of the town. They reportedly did this because on Israeli maps, Ghajar was Lebanese territory, and Israel did not want to involve Lebanon in the war. Ghajar, though, was Syrian—it had been misplaced on 1943 British maps. A delegation from Ghajar, cut off from the rest of Syria during the war, traveled to Beirut to ask to be annexed; Lebanon was not interested. Three months after the war, another delegation traveled to Israel and asked that the village become Israeli. Only then did Israeli control extend north through Ghajar (Khativ 1988; Wolf 1995a). Only the village itself was included, though, and most of its agricultural land remained in Syria. Mekorot engineers did install a three-inch pipe for drinking water for the villagers from the Wazzani Springs, which, although literally a stone's throw from the village, were left under Lebanese control (Wolf 1995a).

Extensive literature exists on the detailed decision-making regarding events before, during, and after the 1967 war. What is noticeable in a search for references to water resources, either as strategic targets, or even as a subject for propaganda by either side, is the almost complete absence of such references. See, for example, Institute for Palestine Studies (1970), which includes almost no mention of water; Brecher (1974), who includes chapters on both "Jordan Waters" and "The Six Day War," but documents no link; and Laqueur (1967, 50), who claims that "[water] was . . . certainly not one of the immediate reasons for hostilities." Stein and Tanter (1980) do not mention water at all.

Boundaries Following the 1967 War

The boundaries following the 1967 war, determined by an unsigned cease-fire agreement, have generally held until very recently (see Figure 4.8). The boundary between Israel and Lebanon, which was not involved in the war, remained the international boundary of 1923; the boundary between Israel and Syria extended well beyond the Armistice Line and demilitarized zones of 1949 to include the plateau of the Golan Heights as far as Quneitra;[10] and the boundary with Jordan returned past the Green Line of 1949 to the 1922 British-Mandate division between Palestine and Transjordan along the Jordan River.

In the territorial gains and improvements in geostrategic positioning which Israel achieved in the war, Israel also improved its "hydrostrategic" position: With the Golan Heights, it now held all of the headwaters of the Jordan, with the exception of a section of the Hasbani, and an overlook over much of the Yarmuk, together making the Headwaters Diversion Project impossible. The West Bank not only provided riparian access to the entire length of the Jordan River, but it overlay three major aquifers, two of which Israel had been tapping into from its side of the Green Line since 1955 (Garbell 1965, 30). Jordan had planned to transport 70–150 mcm/yr from the Yarmuk River to the West Bank. These plans, too, were abandoned.

The War of 1982

In 1982, Israel, for the second time, mounted an operation against the Palestine Liberation Organization (PLO) in Lebanon. The first time, "Operation Litani," four years earlier, Israel had stopped its advance at the Litani River and, before withdrawing, had turned over portions of southern Lebanon to the South Lebanon Army under the command of

FIGURE 4.8. International Borders, 1967 to Present (Redrawn from Wolf 1995a).

Major Sa'ad Haddad. Haddad was reportedly to protect Israeli interests in the region, particularly defending against attempted Palestinian incursions through the area to Israel. In addition, Haddad's militia was reported to have protected the Jordan headwaters of the Hasbani by closing some local wells and preventing the digging of others (Naff and Matson 1984, 49). Some Israelis involved in these issues contest these reports. According to an Israeli officer who dealt with Haddad extensively, the Lebanese major made perfectly clear to the Israelis that "We will cooperate with you, but there are two subjects which are taboo—our land and our water" (Wolf 1995a, 58).

In 1979, engineers from Mekorot, Israel's water planning agency, developed plans to divert 5–10 mcm/yr from the Wazzani Springs for irrigation in Shi'ite southern Lebanon and in Israel. To allow the project to flow on gravity alone, a slight northward modification of the Israeli-Lebanese border, by about one kilometer, was considered (Wolf 1995a, 72). These plans, too, were challenged by Haddad. (Amery [1996, personal communication] points out the power disparity between Haddad and Israeli authorities, and questions whether Haddad's objections would have been sufficient to deter Israeli plans.)

In the 1982 operation, the initially stated objective was a 40-km buffer strip, but, by July, Israeli forces had surrounded Beirut. After the invasion was launched by then–Defense Minister Ariel Sharon, a "water hawk" who had frequently spoken of seizing the Litani, Israel captured the Qir'awn Dam and immediately confiscated all hydrographic charts and technical documents relating to the Litani and its installations (Cooley 1984, 22). Despite Israel's redeployment in 1985, it has retained a "Security Zone" of territory, extending from the international boundary north to a bend in the Litani, which had been established in 1978.

The same absence of documentation linking water and war is true for Israeli reasons for launching operations in Lebanon in 1978 and 1982 (see, for example, MacBride 1983). As noted previously, Israel's ally in southern Lebanon, Major Sa'ad Haddad, had made clear to Israel in 1979 that water was a taboo subject. It was Haddad, too, who objected to Israel's 1979 plans for a diversion of the Wazzani Springs. Avraham Tamir (1988), a major-general who helped outline Israel's strategic needs in 1967 and in 1982, described in detail the military strategy of the 1982 war—again, mention of water is conspicuously absent.

While the Israel Defense Forces planning branch does have an officer whose responsibilities include water resources, both the officer with those responsibilities during the 1982 war and Tamir (personal communications 1991) insist that water was not, even incidentally, a factor in the war. When pressed on the subject, Tamir replied:

Why go to war over water? For the price of one week's fighting, you could build five desalination plants. No loss of life, no international pressure, and a reliable supply you don't have to defend in hostile territory.

Even if water was not the immediate cause of the war, the question remains, Does Litani water reach Israel? Despite the inherent difficulty in proving the absence of something, my answer, after investigating as closely as possible, is "No." This conclusion is based on the following (discussed in more detail in Wolf 1995a, 76–78)[†]:

- The Litani River has a natural flow of about 700 mcm/yr. A dam at Qir'awn in the Beka'a Valley and irrigation and hydropower diversions completed in the mid-1960s reduce the lower Litani flow to 300–400 mcm/yr (Kolars 1992). This lower section, flowing within kilometers of the Hasbani and the Israeli border, historically had presented the possibilities of diversions in conjunction with the Jordan system, and Israel has carried out seismic studies and intelligence reports to determine the feasibility of a Litani diversion (Naff and Matson 1984, 76). These reports concluded that, given the small amount left in the lower Litani, a diversion would be economically unattractive and, in any event, would be politically infeasible until cooperation could be developed with Lebanon (Wolf 1995a, 58). The Lebanese position was and continues to be that rights to Lebanese water should be retained for future Lebanese development.
- Reports of a secret diversion tunnel were investigated by UN forces, as well as by members of the international press, to no avail (Soffer 1994). Satellite photos (LANDSAT and SPOT), air photos (Israel Air Force), Mekorot maps, and field investigations (June 1987; June, October, December 1991), all show only two water pipelines crossing the Lebanon-Israel border—a three-inch pipe from the Wazzani Springs in Lebanese territory to the town of Ghajar in Israel, previously mentioned, and a ten-inch pipe from Israel into the Lebanese village of R'meish.
- Hydrologic records show neither any unaccounted-for water in the Israeli water budget after 1978, nor any increases in the average flows of the Ayoun or the Hasbani, the most likely carrier streams for a diversion. Because of three years of drought, on 14 October 1991, the Israeli Water Commissioner asked the Knesset to allow pumping of the Sea of Galilee below the legal "Red Line," the legal water level below which the entire lake is in danger of becoming saline. On the same day, a field investigation showed that both the Ayoun and the Hasbani above the Wazzani Springs were dry.
- A hypothetical trucking operation is even more difficult to prove or disprove. Both officials in Mekorot (Turgemon, personal communication, October 1991) and Israeli officers responsible for southern Lebanon acknowledge that witnesses may have seen Israeli military water trucks in southern Lebanon. Each has suggested that the most

likely explanation is that the trucks were carrying drinking water from Israel for Israeli troops stationed in the "Security Zone." The Israeli military code, they point out, insists that soldiers drink water only from official collection points, all of which are in Israel.

An officer who has acted as liaison between Israeli and South Lebanon forces doubts that anyone saw Israeli trucks filling at the Litani, pointing out that the 20-ton "Rios" which are used to carry water could not make the grade of the military road leading away from the Litani if the trucks were full (Wolf 1995a). Soffer (1994, 7) has calculated that a cubic meter of water trucked from the Litani into Israel would cost about $4 to $10, as compared to about $1.50 for a cubic meter of desalinated water.

Water and War—Conclusions

After closely examining the arguments both in favor and against a "hydrostrategic" imperative driving military and territorial decision-making, it is possible to answer the question posed at the beginning of this section: "Has territory been explicitly targeted, captured, or retained because of its access to water resources?" I argue for the following conclusions:

- Water resources were not a factor for strategic planning in the hostilities of 1948, 1967, 1978, or 1982. By this I mean that the decision to go to war, and strategic decisions made during the fighting, including which territory was necessary to capture, were not influenced by water scarcity or the location of water resources. The locations of water resources were not considered strategic positions (except in the purely military sense), nor were they a factor in retaining territory immediately after the hostilities.
- There is no evidence that Israel is diverting any water from the Litani River, either by pipe or by truck. In fact, since 1985, when central southern Lebanon lost its own water supply, an average of 50,000 m^3/month has been piped into that region from wells in northern Israel (Wolf 1995a, 59). As a consequence, it is difficult to view water as a rationale for Israeli retention of Lebanese territory.[11]

WATER AND NEGOTIATIONS

As noted above, the years of warfare have obscured almost all of the original boundaries between Israel and her neighbors: The June 1967 war erased the boundaries determined in the 1949 armistice agreement on all fronts, with the exception of the line between Israel and Lebanon, in favor of unsigned cease-fire lines between the combatants. These in turn gave way to the lines of the disengagement agreement of 1974 on the fronts between Israel and Syria, and Israel and Egypt. The post-1967 years, and particularly the recent period of peace talks, have been characterized by a quest for stable boundary lines, taking into account the lessons of the past. The question I seek to answer in this section is, "How much of the quest for permanent boundaries is influenced by the location of water resources?"

Boundary Proposals: Strategy and Hydrostrategy

The search for acceptable boundaries began immediately after the June 1967 war. For Israel, the guiding rationale was that territorial concessions should be balanced by security needs, defined differently depending on where one was on the political spectrum—from retaining to relinquishing all of the captured territory. For the Arabs, regaining *all* land captured during the war became the operative imperative, shadowing subsequent negotiations. In the survey of boundary proposals which follows, I exclude the extreme positions of either side—from a "Greater Israel" on one side to an Israel which ceases to exist on the other—but rather describe those which incorporate the concept of territorial compromise in exchange for peace, as provided for in United Nations Resolutions 242 (1967) and 338 (1973).[12] Since Israel controls the territory, and presumably would not withdraw unless its strategic and political goals were met in negotiations, the analysis focuses on studies which investigate Israeli interests in withdrawal.[13] However, Foucher's (1987) warning should be borne in mind, that, "in the case of the strategic debate over Israel, the West Bank and neighboring Arab states, one may have reason to consider that 'secure borders' for Israel and 'security for all' are not synonymous concepts."

Immediately after the 1967 war, strategic needs, none of which related to water, were spelled out by the Israeli government, which, if met, would result in Israeli withdrawal from occupied territory. According to Moshe Dayan, the Golan Heights were negotiable even without a peace treaty, and, with such a treaty, so was the rest of the territory captured in 1967, except East Jerusalem (Slater 1991, 286–290). This approach was met by the "three noes" of an Arab Summit in Khartoum in August 1967—no peace, no recognition, and no negotiations with Israel

(Sachar 1979, 676). Israel then strengthened its position in the newly occupied territories through settlement activities: a string of kibbutzim was established on the Golan Heights in 1967–1968 from Senir, on Givat Banias overlooking the Banias headwaters, south along the ridge of the Heights overlooking the previously demilitarized zone, to Mevo Hama, adjacent to Hammat Gader with its access to the Yarmuk River. Senir and Hammat Gader were each situated in what had been until the war the demilitarized zone—territory which had been part of the British Mandate, but that Syria had occupied in the 1948 war, then had withdrawn from under demilitarized conditions as part of the 1949 armistice agreement.

The same strategy of holding conquered land as inducement to peace talks was followed immediately after the 1982 war in Lebanon. In 1983, an Israeli-Lebanese agreement was signed (but not ratified) which called for an Israeli withdrawal from all of Lebanon. The agreement was abrogated in 1984, however, and, consequently, Israel justifies its continued presence in the "Security Zone" (Tamir 1988).

The Allon Plan

As mentioned, Israel began to use settlement activity as a way of reinforcing its strategic interests immediately following the war of 1967. The Labor government which ruled Israel until 1977 adhered generally to guidelines devised by minister Yigal Allon primarily to address Israeli security concerns (see Figure 4.9). Emphasis was placed mainly on the Jordan Valley and the eastern slope of the West Bank mountain range facing Jordan. It should be noted that the Allon Plan was never endorsed or ratified by any Israeli government, although it guided the Labor government's settlement policy until 1977.

Some Israeli settlements were established outside of the Allon proposals during the Labor years—a few reportedly to help protect Israel's groundwater resources on the northwest corner of the West Bank. As mentioned, Israel had been tapping into the Western Mountain Aquifer, which originates on the West Bank, since 1955. Because of the disparate depths to water for this aquifer in the coastal plain and in the Judean hills (about 60 m in the plain, 150–200 m in the foothills, and 700–800 m in the hills) (Goldschmidt and Jacobs 1958; Weinberger 1991), and the resulting cost differences in drilling and pumping wells in these areas, this aquifer is especially vulnerable to overpumping along a narrow westernmost band of the northern lobe of the West Bank, in the region of Kalkilya and Tulkarm (see Figure 4.10). Some settlement plans

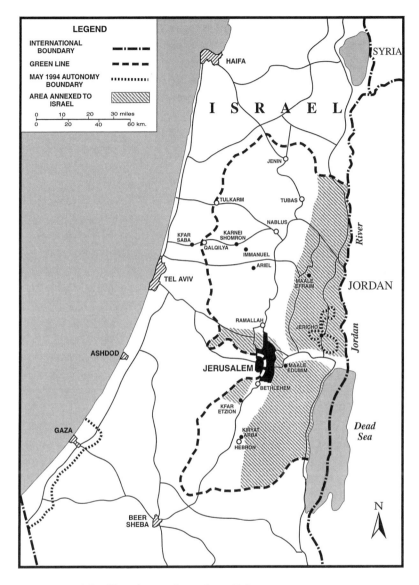

FIGURE 4.9. The Allon Plan (Redrawn from Alpher 1994).

for the late 1970s referred in part to this line, and about five settlements around Elkanna were reportedly sited in part to guarantee continued Israeli control of the water resources on its side of what would soon be referred to as a "red line" (Pedhatzor 1989; State of Israel memoranda, April–July 1977).

FIGURE 4.10. West Bank Groundwater (Redrawn from Shuval 1992).

Post-1977 Boundary Studies

In 1977, the right-wing Likud Party gained control of the Israeli parliament for the first time. As Israeli Prime Minister Menachem Begin was preparing for negotiations with Egyptian President Anwar Sadat, he asked then–Water Commissioner Menachem Cantor to provide him with a map of Israeli water usage from water originating on the West Bank, and to provide guidelines to where Israel might relinquish control, if protecting Israel's water resources were the only consideration.

As described above, Cantor concluded that a "red line" could be drawn, beyond which Israel should not relinquish control, north to

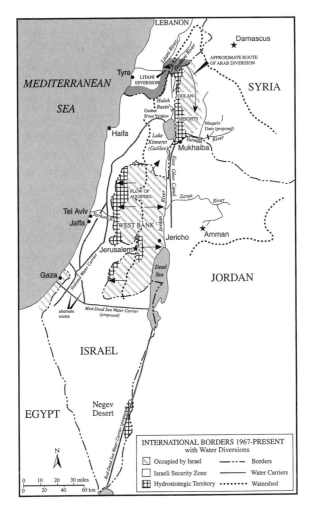

FIGURE 4.11. Hydrostrategic Territory (Redrawn from Wolf 1995a).

south following roughly the 100–200 m contour line along both "lobes" of the West Bank (see Figure 4.11). Israeli water planners still refer to this "red line" as a frame of reference (Wolf 1995a), and it has occasionally been included in academic boundary studies of the region. This concept was later expanded by others to areas of the northern headwaters and the Golan Heights (see Figure 4.11). Brief descriptions of those studies which mention water as a territorial imperative follow.

Cohen's "Defensible Borders"

Saul Cohen (1986) explored "the geopolitics of Israel's border question," addressing possible boundary negotiations with a Palestinian political entity, and with Syria over the Golan Heights. His recommendations for boundary adjustments were considered from the perspective of defensible borders for Israel within the framework of territorial compromise, and included factors of a "strategic-tactical" and a "demographic-economic" nature. They included, explicitly, defensive depth, surveillance points, marshaling areas and corridors, water control, space for Israeli population and industrial growth, absence of dense Arab populations, and psycho-tactical space (1986, 4).

In illustrating the influence of the above principles, Cohen described how water might influence territorial compromise (1986, 55): Israel would need to retain sovereignty over the Banias–Har Dov–Hermon shoulder headwaters region, the Golan slopes east of the Upper Jordan, and the Golan Heights that overlook the Sea of Galilee and the Lower Yarmuk and its Raqqad tributary. On the West Bank, Cohen argued that Israel should annex the territory which extends until the "subterranean water divide," which he identifies as extending from 2 to 6 km east of the Green Line.[14] Despite his acknowledgment that this territory includes Arab population centers, although none of the larger towns, Cohen argued that the substantial geopolitical advantage that Israel would gain (presumably over and above the hydrostrategic considerations) would outweigh those concerns. Overall, for all the factors listed above, Cohen advised Israel to annex approximately 20 percent of the West Bank, 19 percent of the Gaza Strip, and 50 percent of the Golan Heights (1986, 4; see also Figure 4.12).

The Jaffee Center's "Arrangements"

In 1991, the Jaffee Center for Strategic Studies of Tel Aviv University asked two researchers, Yehoshua Schwartz, the director of Tahal, Israel's water planning agency, and Aharon Zohar, also at Tahal at the time, to undertake a study of the regional hydrostrategic situation and the potential for regional cooperation. The result, a three-hundred-page document titled *Water in the Middle East: Solutions to Water Problems in the Context of Arrangements between Israel and the Arabs,* was one of the most comprehensive studies of its kind (Schwartz and Zohar 1991). It examined a number of possible scenarios for regional water development, including possible arrangement between Israel and Jordan, Syria,

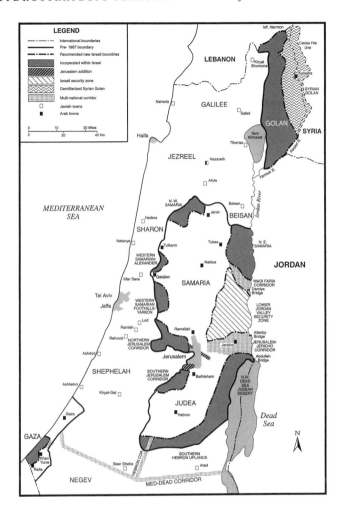

FIGURE 4.12. Cohen's Defensible Borders (Redrawn from Cohen 1986).

Lebanon, Egypt, Turkey, Saudi Arabia, Iraq, and the Palestinians on the
West Bank and Gaza. Scenarios were included both for regional cooperation and for its absence. Evaluations included hydrologic, political,
legal, and ideological constraints. The impacts of potential global climatic change were also considered. The study showed, in the words of
Joseph Alpher, the director of the Jaffee Center, "the potential beauty of
multi-lateral negotiations" (Wolf 1995a).

Some of the findings of the study contradicted government policies at
the time, however. In the sections on possible arrangements between Is-

rael and the Palestinians, and between Israel and Syria, maps of the West Bank and Golan Heights included lines to which Israel might relinquish control of the water resources in each area without overly endangering its own water supply. The line in the West Bank, which was based on Cantor's "red line," suggested that Israel might, with legal and political guarantees, turn control of the water resources of more than two-thirds of the West Bank over to Palestinian authorities without threatening Israel's water sources from the Yarkon-Taninim (Western Mountain) aquifer, although the authors advocated retaining control beyond the "red line." The same was true of more than half of the Golan Heights (see Figure 4.11).

These maps contradicted the position of the Ministry of Agriculture. Headed by Rafael Eitan of the right-wing Tzomet Party, the ministry's position was that, to protect Israel from threats to both its water quantity and quality, Israel had to retain political control over the entire West Bank.[15] On 12 December 1991, seventy copies of the report were sent throughout Israel for review, including to the Ministry of Agriculture. Calling the maps mentioned "an outline for retreat," Rafael Eitan and Dan Zaslavsky, whom Eitan had recently appointed Water Commissioner, insisted on a recall of the review copies and a delay in the release of the report. In January 1992, the Israeli military censor backed the position of the Ministry of Agriculture and, citing the sensitivity of the report's findings, censored the report in its entirety.[16]

Alpher's Proposals

Once negotiations began in 1991, as explored below, boundary proposals took on new urgency. In one of the most comprehensive postnegotiation examinations of West Bank boundary options, Joseph Alpher (1994) both summarizes previous proposals for final boundary arrangements, including the recent "Third Way" and Sharon plans (neither of which has a hydrostrategic component), and offers his own. In defining Israel's requirements in a negotiated agreement, he characterizes Israel's needs according to nine parameters: security, water, demography and politics, the heritage dimension, the historic dimension; Hebron; Israeli Arabs and the danger of irredentism; the economic dimension; and the need to straighten the borderline (boundary).

Alpher relies on the unpublished Jaffee Center report of 1991 described above for his description of hydrostrategic territory.[17] He delineates West Bank territory which might be annexed to Israel in order to protect the Western Mountain Aquifer, as defined first by Cantor's "red

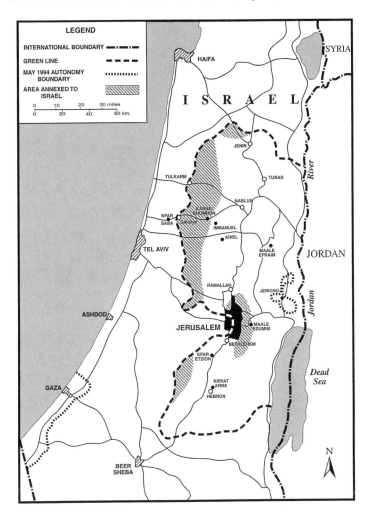

FIGURE 4.13. Alpher's Water Boundaries (Redrawn from Alpher 1994).

line"—the westernmost section of the northern lobe of the West Bank, and a region around Jerusalem (see Figure 4.13). Alpher notes that, while annexation would guarantee Israeli control of the water resources, adequate supervision and control arrangements are possible without annexation—perhaps through the implementation of a joint water regime (1994, 28). He also notes that the territories of the West Bank which are vital for continued control over water management have already been heavily settled because of their importance with regard to security. Thus, he concludes:

FIGURE 4.14. Alpher's Final Boundaries (Redrawn from Alpher 1994).

. . . the water issue is not necessarily a decisive rationale for annexation. At the same time, to the extent that the water issue is juxtaposed geographically with additional vital issues such as security and demography, then it may be seen to further enhance an annexation solution. (1994, 28)

Alpher finally seems to weigh in against annexation. In his final map incorporating all of the parameters he defines as crucial, *no* territory which was identified as being important for water alone is slated for annexation (see Figure 4.14).

Bilateral and Multilateral Negotiations

The Gulf War in 1990 and the collapse of the Soviet Union caused a realignment of political alliances in the Mideast which finally made possible the first public face-to-face peace talks between Arabs and Israelis, in Madrid on 30 October 1991. During the bilateral negotiations between Israel and each of its neighbors, it was agreed that a second track be established for multilateral negotiations on five subjects deemed "regional," including water resources. These two mutually reinforcing tracks—the bilateral and multilateral—have led, at this writing, to a treaty of peace between Israel and Jordan, and a declaration of principles for agreement between Israel and the Palestine Authority. Both have had a water component in terms of allocations and projects. In *neither* has water had any influence on the discussions over final boundaries.

Israel-Jordan Treaty of Peace

Israel and Jordan have had probably the warmest relations of any two states legally at war. Communication between the two has taken place since the creation of each, ameliorating conflict and facilitating conflict resolution on a variety of subjects, including water. As noted above, the "Picnic Table Talks" on allocations of the Yarmuk have taken place since the 1950s, and negotiations formulating principles for water-sharing projects and allocations have occurred in conjunction with, and parallel to, both the bilateral and multilateral peace negotiations.[18] These principles were formalized on 26 October 1994, when Israel and Jordan signed a treaty of peace, ending more than four decades of a legal, when not actual, state of war.[19]

For the first time since the states came into being, the treaty legally defines mutually recognized water allocations. Acknowledging that "water issues along their entire boundary must be dealt with in their totality," the treaty spells out allocations for both the Yarmuk and Jordan Rivers and Araba/Arava groundwater, and calls for joint efforts to prevent water pollution. Also, "[recognizing] that their water resources are not sufficient to meet their needs," the treaty calls for ways of alleviating the water shortage through cooperative projects, both regional and international.

The peace treaty also makes some minor boundary modifications. As noted, the Israel-Jordan boundary was delineated by Great Britain in 1922, and followed the center of the Yarmuk and Jordan Rivers, the Dead Sea, and Wadi Araba/Arava. In the late 1960s and 1970s, Israel had occasionally made minor modifications in the boundary south of the Dead Sea to make specific sections more secure from infiltrators. It had also done so on occasion to reach sites from which small wells might better be developed. Since at least 1980, no modifications were made except on the rare occasion that one of these local wells ran dry and had to be redug. *All* of these territorial modifications were reversed and all affected land was returned to Jordan as a consequence of the peace treaty, although Israel retains rights to the water which comes from these wells. Moreover, a small enclave of Jordanian territory in the Wadi Araba/Arava is being leased back to Israel in twenty-five-year increments.

One other area was similarly affected. In 1926, a Jewish entrepreneur named Pinhas Rutenberg was granted a seventy-year concession for hydropower generation at the confluence of the Yarmuk and Jordan Rivers on land leased by Transjordan. The dam he built for that purpose was destroyed in the fighting of 1948, and the 1949 Armistice Line left a small portion of Jordan under Israeli control. This land was farmed by the kibbutz Ashdot Ya'akov, which was established in 1933. With the 1994 peace treaty, sovereignty over the land was returned to Jordan, which in turn leased it back to the Israelis—Israeli kibbutznikim now travel into Jordanian territory regularly to farm their land.

Israeli-Palestinian Declaration of Principles and Interim Agreement

On 15 September 1993, the "Declaration of Principles on Interim Self-Government Arrangements" was signed between Palestinians and Israelis, calling for Palestinian autonomy in, and the removal of Israeli military forces from, Gaza and Jericho. Among other provisions, this bilateral agreement called for the creation of a Palestinian Water Administration Authority. Moreover, the first item in Annex III, on cooperation in economic and development programs, included a focus on:

> Cooperation in the field of water, including a Water Development Program prepared by experts from both sides, which will also specify the mode of cooperation in the management of water resources in the West Bank and Gaza Strip, and will include proposals for studies and plans on water rights of each party, as well as on the equitable utilization of joint water resources for implementation in and beyond the interim period.

At approximately the same time, Israeli water managers discovered an additional 70 mcm/yr of available yield in the Eastern Mountain aquifer—the only of the three main West Bank units which was not being overpumped at the time. This probably did not hurt Jericho's choice as the first West Bank town to be given autonomy.[20]

Between 1993 and 1995, Israeli and Palestinian representatives negotiated to broaden the interim agreement to encompass greater West Bank territory. On 28 September 1995, the "Israeli-Palestinian Interim Agreement on the West Bank and the Gaza Strip," commonly referred to as "Oslo II," was signed in Washington, D.C. The question of water rights was one of the most difficult to negotiate, with a final agreement postponed and to be included in the negotiations for final status arrangements.[21] Nevertheless, tremendous compromise was achieved between the two sides: Israel recognized the Palestinian claim to water rights, of an amount to be determined in final status negotiations, and a Joint Water Committee was established to cooperatively manage West Bank water and to develop new supplies. This committee also was to supervise joint patrols to investigate illegal water withdrawals—its first "action" was to discover and put a stop to illegal drilling in the area of Jenin in December 1995 (Israel Line, 20 December 1995).

According to the agreement, Israeli forces are scheduled to withdraw from 6 Palestinian cities in order from north to south, and from 450 towns and villages throughout the West Bank. The final status of Israeli settlements in the West Bank has yet to be determined. *No territory whatsoever was identified as being necessary for Israeli annexation due to access to water resources.* The second and third cities scheduled for Israeli withdrawal—Tulkarm and Kalkilya, fall well within the "red line" delineated in Israeli studies as being necessary to retain for water security.

Negotiations between Israel, Syria, and Lebanon

As of this writing, water has not been raised in official negotiations between Israel and Syria.[22] Serious bilateral negotiations have only taken place since the fall of 1995, and, given the influence Damascus has on Beirut, Israel-Lebanon talks are not likely until Israel and Syria make more progress. Israelis had hoped to begin talks on water resources with the Syrians at a meeting in Maryland in January 1996, but the Syrians reportedly refused to broaden the scope (Israel Line, 24 January 1996).

The premise of Israel-Syria negotiations is an exchange of the Golan Heights for peace. The discussions thus far have focused on interpretations of how much Golan, and with what security arrangements, for how much peace. The crux of the territorial dispute is the question of to which boundaries Israel would withdraw—the boundaries between Israel and Syria have included the international boundary between the British and French Mandates (1923), the Armistice Line (1949), and the cease-fire lines from 1967 and 1974.

The Syrian position has been an insistence on a return to the borders of 5 June 1967, while Israel refers to the boundaries of 1923. Although it has not been mentioned explicitly, the difference between these two positions is precisely over access to water resources. The only distinction between the two lines is the inclusion or exclusion of the three small areas which made up the demilitarized zone between 1949 and 1967—Givat Banias, the hill overlooking Banias Springs; the Daughters of Jacob Bridge area; and the town of El-Hama/Hammat Gader—a total of about 60 km². These three territories were included in British Palestine specifically because of their access to the Jordan and Yarmuk Rivers, and, since each is a relatively low-lying area with no strategic importance,[23] their access to water is still considered paramount.

In fact, even before Israel-Syria negotiations began, a flurry of articles had stressed the importance of water on the Golan Heights. As mentioned above, Schwartz and Zohar (1991) advised Israeli retention of the Golan Heights west of the Jordan River watershed line in order to guarantee continued control of both water quantity and quality (see Figure 4.11). In a 1994 study, Aryeh Shalev (1994), himself a retired general in the Israeli army, cited five other retired generals on the importance of Israeli sovereignty over the Golan to the protection of water resources. Even in his small sample, Shalev found a spectrum of opinion, from Major-General (res.) Hofi, who suggested that Israel needed to retain a physical presence on the Golan Heights, to Major-General (res.) Shafir, who advocated retention of at least the plateau above the Sea of Galilee, to former chief of staff Gur, who concluded that the water problem could be resolved politically in a peace treaty, and that the territory was not vital. Shalev concluded that Syria would not risk a war with Israel for water, especially since a diversion would take years to construct and would constitute a clear *casus belli*. It stands to reason, Shalev argued, that countries involved in water-sharing agreements would want to maintain them.

In the meantime, Schiff (1995), Tarnopolsky (1996), and others have argued in the popular Israeli and Jewish press that water's paramount importance may scuttle negotiations over the Golan, while Israeli politicians from the ruling Labor Party, including Prime Minister Shimon Peres and his Foreign Minister Ehud Barak, argue that while the land may be negotiable, the water is not (*Jerusalem Post,* 6 January 1996 and 27 January 1996).

Water and Negotiations—Conclusions

In answer to the question posed at the beginning of this section—"How much of the quest for negotiated boundaries has been influenced by the location of water resources?"—the evidence seems to suggest: not much. This is not to say that water has not been an important topic in each set of negotiations—quite the opposite is true. The questions of water allocations and rights have been intricate and are being resolved with great difficulty. Nevertheless, with the concluded negotiations between Israel and Jordan, and the ongoing talks between Israel and the Palestinians, and despite the quantity of studies identifying hydrostrategic territory and advising its retention, *no* territory to date has been retained simply because of the location of water. Solutions in each case have focused on creative joint management of the resource, rather than insistence on sovereignty.

The pattern which does seem to be emerging, however, is that water, *in addition* to one or more other concerns, may justify retention of territory. For example, in the absence of any legal claims, security interests, or settlements, Israel withdrew from all Jordanian territory which it had occupied—even those small portions which had hydrostrategic importance. What was important was an agreement on water management, not territory.

In contrast, Israel developed some settlements on the westernmost side of the West Bank to address security and demographic concerns in addition to protecting its water supply. While sovereignty over most of this territory is being turned over to the Palestinians, determination of the final status of those settlements has been postponed until the final round of negotiations. Again, the solution was found through agreements guaranteeing joint water management of shared aquifers, precluding the need for annexation of territory.

These principles may be played out in negotiations between Israel and Syria as well. While Syria argues for the Armistice Line as it stood on 5 June 1967, Israel insists that boundaries be based on the 1923 international division between the British and French Mandates—the difference being three small areas of vital hydrostrategic importance. Based on the experience of other boundary delineations (including those between Israel and Egypt, and Israel and Jordan), Israel will probably prevail in a legal sense (the Armistice Lines were explicitly temporary). Thus, Israel is able to address hydrostrategic territory *in conjunction* with its legal claims.

CONCLUSIONS

The conclusions which might be drawn from this chapter have implications beyond interpretations of the Arab-Israeli conflict. The first are those for the importance of historical analysis in deriving these interpretations. As Michel Foucher (1989) describes cross-border "realities and representations":

> Border analysis does not only deal with space, but also with time. So boundaries can be considered as *time written in space*—not merely the past, but also a special relationship with that past which stands as an inevitable background to current geopolitical decisions.

The question I have sought to address in this chapter examines the existence of hydrostrategic territory—that is, territory over which sovereignty has been sought politically or militarily *solely* because of its access to water sources, and in the absence of any other compelling strategic or legal rationale—and its role in boundaries, warfare, and negotiations. The answer seems to be a qualified "yes" that hydrostrategic territory has existed as a political goal; a definite "no" for water-related sites as military targets; and a convoluted "maybe" for hydrostrategy as a focus of peace negotiations. The question was then divided in three to better define the issues:

1.   Have boundaries been drawn historically on the basis of the location of water access? Beginning with the Paris Peace Talks in 1919 and ending with the 1923 Mandate boundaries, the Zionists clearly defined their future state in hydrologic terms, seeking as much of the Jordan basin as possible, and occasionally some of the Litani as well. In these goals they were only marginally successful, losing two of the three headwaters, but retaining most of the flow of the Upper Jordan and all of the Sea of Galilee. The watershed was further divided in the 1922 creation of Transjordan in the territory east of the center of the Jordan River, Dead Sea, and Wadi Araba/Arava. The Zionists attempted to reinforce their sovereignty over the headwaters region through settlement activity in the late 1930s, always within negotiated boundaries.

It seems clear that water was uppermost in the minds of planners and political decision-makers, particularly Zionists, as boundaries were negotiated over the years, at times assuming importance equal to more traditional definitions of security, and that specific territory was sought for its access to the water resources alone. However, despite studies advocating the need for greater access to water through 1947, actual official advocacy of sovereignty over such hydrostrategic territory ceased each and every time negotiations over legal borders were concluded.

2. During warfare between competing riparians, has territory been explicitly targeted, captured, or retained because of its access to water sources? This has been the most elaborately argued question in the literature relating water resources to Arab-Israeli relations, although it is too rarely investigated in detail. Contradicting the views of functionalists and advocates of a "hydrostrategic imperative," water-related territory seems actually to have played almost no role at all in Arab-Israeli warfare. Close examination of strategic planning and military decision-making and tactics suggests no evidence at all that water was a factor in the hostilities of 1948, 1967, 1978, or 1982. The *only* instance in which territory was sought for access to water was a brief attempt by Israel in 1979 to move the boundary with Lebanon about a kilometer north to gain access to the Wazzani Springs—an attempt objected to by the local Lebanese commander and never implemented.

3. In the course of peace negotiations, has hydrostrategic territory been seen as vital to retain by any of the riparians? Here, too, the answer seems to be that, despite a flurry of studies recommending Israeli retention of territory to protect its water sources, *no* territory to date has been retained simply because of the location of water. Before and during each set of concluded negotiations, both popular and academic commentary has appeared arguing that territory is critical for hydrostrategy, yet the actual solutions in each case have focused on creative joint management of the resource, rather than insistence on sovereignty. That is not to say that water has not been a difficult topic for negotiations between Israel and its neighbors—quite the opposite is true, but the debate has been over rights, allocations, and management, *not* over territory. Of territory identified in Israeli studies as being vital to the protection of Israel's water resources, *none* has been retained by Israel simply because of the location of

water alone. This has been true of agreements completed as of this writing—the 1994 treaty of peace between Israel and Jordan, and the 1993 Declaration of Principles and the 1995 Interim Agreement between Israel and the Palestinians—where arrangements were made for joint management, in lieu of sovereignty. Such hydrostrategic territory *is* being insisted on, however, in ongoing negotiations *only* when at least one other compelling justification exists. Israel may point to demographic and/or security concerns over some Jewish settlements which were reportedly sited to protect Israel's groundwater resources, for example. And Israel's insistence that boundaries with Syria be drawn at the 1923 Mandate line rather than according to the temporary 1949 armistice agreement is based on precedent and international law as well as on hydrostrategic needs.

A close examination of the historic facts seems to show that water has had much less impact on the Arab-Israeli conflict than is increasingly argued, certainly in strategic, spatial, and territorial terms. As Libiszewski (1995, 93) concludes in a thorough study of water and security in the Middle East, the Arab-Israeli conflict "is not primarily a struggle 'over water.' The conflict is over national identity and existence, territory, as well as over power and national security." In this context, water has played a minor role, but only, it seems, in conjunction with one or more of these overriding imperatives.

The second set of implications of this chapter is for those proponents of the existence of "water wars," both within the region and within the broader context of "environmental security." If the link between water and armed conflict can be so overemphasized so commonly in this "worst-case scenario," should we not examine more closely the causal argument of resource conflicts in general? The more interesting question seems *not* to be, Where will the next "water war" break out? But rather, given such a vital and scarce resource which flows so often between hostile riparians, What is it about water which seems to guide those dependent on it away from conflict, and toward cooperation? The true lesson of the Arab-Israeli experience seems not to be of water as exacerbator of conflict, but rather, as the people in the region move from war to peace and the desire for sovereignty gives way to principles of joint management, of water as inducer to cooperation. As Lord Curzon (1907, cited in Prescott 1987, 5) has said, "Frontiers are indeed the razor's edge on which hang suspended the modern issues of war and peace." It is perhaps this latter aspect that needs to be more stressed.

NOTES

Author's Acknowledgments: I am grateful to Bruce Maddy-Weitzman and the Dayan Center for Middle Eastern and African Studies at Tel Aviv University for hosting me so graciously. A shorter version of this chapter was presented at the conference Transformations of Middle Eastern Environments: Legacies and Lessons, Yale University, 28 October–1 November 1997, and published in the conference proceedings (*Yale School of Forestry and Environmental Studies Bulletin*, Series #103, 1998). Some portions, particularly those regarding early history and the "hydraulic imperative," are drawn from Wolf 1995a.

    1. "Hydraulic" refers to the mechanics of water under pressure. "Hydrostrategic" describes the link between the location of water resources and strategic decision-making.

    2. The reader interested in overviews of the water shortage is referred to Lowi (1993); Hillel (1994); Isaac and Shuval (1994); Kliot (1994); Lonergan and Brooks (1994); Allan and Mallat (1995); Murakami (1995); and Wolf (1995a). On technical aspects, annotated bibliographies in support of the Middle East water multilateral negotiations were undertaken by the U.S. Agency for International Development (1992), the International Water Engineering Centre at the University of Ottawa (1993), and SOAS, the University of London (1995). In 1996, the national academies of science of the United States, Israel, Palestine, and Jordan began a multiyear study of water policy and technology appropriate for the Middle East. Their report is due in late 1999.

    3. "You will flounder if you like, but you will not flounder at our expense."

    4. The boundary was originally defined as the location of the river in 1922. However, after the Jordan shifted 800 meters westward during the winter of 1927–1928, the British high commissioner decided that the boundary would henceforth be the center of the water bodies mentioned, wherever they meandered. A similar shift in the streambed, and consequently in the boundary, occurred in the winter of 1978–1979 (Biger 1994).

    5. "Hydraulic" refers to the mechanics of water under pressure. "Hydrostrategic" describes the link between the location of water resources and strategic decision-making.

    6. This last, most extreme, scenario is described in detail in Stauffer (1982).

    7. Here, too, Beaumont is mistaken—Lebanon was not involved in the 1967 war.

    8. The eighth point of a British Eight-Point Plan submitted in July 1949 was a call for "an Israeli-Arab agreement for sharing the waters of the Jordan and Yarmuk Rivers" (Caplan 1993, 87).

    9. At that scale, the width of the line alone allowed for territorial ambiguity of 250 meters along the boundary (Biger 1989).

    10. There was some change along the Israel-Syria border as a consequence

of the 1974 war. During that war, Syrian forces crossed westward past Quneitra but did not descend from the Golan Heights. Although the Israeli counterattack extended east again across Quneitra, the town itself was returned to Syria following a 1974 disengagement agreement (Hareven 1977; Sachar 1979).

11. Nevertheless, Amery (1993) has speculated that Israel will, at a minimum, pressure Lebanon to make Litani water available as a prerequisite to Israeli military withdrawal.

12. These UN resolutions, the basis for peace negotiations between Israel and each of its neighbors, call for "the right of every State in the area to live in peace within secure and recognized boundaries," and "withdrawal of armed forces from territories occupied." According to one interpretation, definite articles were purposely omitted in the latter clause (i.e., "territories," not "the territories"), specifically allowing for some flexibility in boundary negotiations (see Julius Stone's foreword to Blum 1971).

13. I was not able to find many sources from an Arab perspective which dealt with the specifics of territorial compromise. Two exceptions are Kharmi (1994), who allows for Palestinian territorial adjustments in negotiations *provided* Israel reciprocate with a like amount of land from within Green Line Israel, and Falah (1995), who, with Newman, argues that a "good" boundary between Palestine and Israel would incorporate both internal and external perceptions of threat. Since neither study has an explicit hydrologic component, neither is described in detail here.

14. Cohen is probably referring to Cantor's "red line." The watershed divide is actually several kilometers inland along the ridge of the Samarian and Judean hills.

15. Eitan's position, argued in full-page ads in the Israeli press, has little foundation in hydrogeology, as discussed in Wolf 1995a, 78–80.

16. When peace talks began in 1991, the document remained censored for fear its release would reveal Israeli negotiating strategy. To date, the document has not been released.

17. Alpher was director of the Jaffee Center when it commissioned the 1991 study.

18. For more details on the bilateral and multilateral talks on water, see Wolf 1995b.

19. To my knowledge, these are the first international boundaries defined legally by Universal Transverse Mercator (UTM) coordinates as measured using the Global Positioning System.

20. There is *no* evidence at all the water was even considered in this choice. This comment is only the author's speculation.

21. "Oslo II" estimates the future needs of West Bank Palestinians to be 70–80 mcm/yr. Until a final arrangement is negotiated, the two sides agree to cooperate to find a total of 28.6 mcm/yr for the interim period.

22. In unofficial "Track II" discussions, water was the focus of meetings

where Israelis and Lebanese were present as early as 1993, and where Israelis and Syrians participated in 1994 and in 1997. Participants at these meetings did not necessarily have any official standing.

23. One might argue that the hot springs at Hammat Gader offer economic benefits, but these are relatively minor.

†*Postscript:* Subsequent to our writings, each of us made attempts to clarify perceived ambiguities in the other's arguments vis à vis a possible Litani diversion. Wolf provided Amery with an August 1988 Israeli Foreign Ministry report entitled "Regional Water Resources in the Middle East," which states that the Litani is not an international river and is not diverted into Israel; also offered were primary flow data from the Israel Hydrologic Service for the Hasbani and Ayoun headwaters, which show a decrease in flow from the Hasbani, not exhibited in the other headwaters, from around 1979. Amery argues that no data are provided in the former document to back its conclusions, and the handwritten data in the latter show enough temporal fluctuation to mask a possible small-scale diversion. He argues that the possibility of a water diversion should not be dismissed, especially given continued reports of such, the most recent being from two southern Lebanese members of parliament who reported it to the Lebanese prime minister (Ha'aretz 23 June 1999). While neither of us is convinced of the other's argument, note that more documentation from both countries exists than is mentioned in our respective chapters.

REFERENCES

Allan, J. A., and Chibli Mallat, eds. 1995. *Water in the Middle East: Legal, Political, and Commercial Implications.* London and New York: Tauris Academic Studies.

Alpher, Joseph. 1994. *Settlements and Borders.* Final Status Issues: Israel-Palestinians, Study No. 3. Tel Aviv: Jaffee Center for Strategic Studies.

Amery, H. A. 1993. "The Litani River." *Geographical Review* 83 (3): 230–237.

———. 1998. "The Role of Water Resources in the Evolution of the Lebanon-Palestine Border." *GeoJournal* 44 (1): 19–33.

Beaumont, Peter. 1991. "Transboundary Water Disputes in the Middle East." Submitted at a conference on Transboundary Waters in the Middle East, Ankara, September.

———. 1994. "The Myth of Water Wars and the Future of Irrigated Agriculture in the Middle East." *Water Resources Development* 10 (1): 9–22.

Biger, Gideon. 1989. "Geographical and Other Arguments in Delimitation in the Boundaries of British Palestine." In Conference Proceedings: International Boundaries and Boundary Conflict Resolution, Durham, England, 14–17 September.

———. 1994. "River and Lake Boundaries in Israel." In Clive Schofield and Richard Schofield (eds.), *World Boundaries Vol. 2: The Middle East and North*

Africa, pp. 101–112. World Boundary Series, G. Blake, series editor. London and New York: Routledge.

Bingham, G., A. Wolf, and T. Wohlgenant. 1994. *Resolving Water Disputes: Conflict and Cooperation in the U.S., Asia, and the Near East.* Washington, D.C.: U.S. Agency for International Development.

Biswas, Asit. 1992. "Indus Water Treaty: The Negotiating Process." *Water International* 17 (4): 201–209.

Blum, Yehuda. 1971. *Secure Boundaries and Middle East Peace.* Jerusalem: Hebrew University of Jerusalem Faculty of Law and Institute for Legislative Research and Comparative Law.

Brawer, Moshe. 1968. "The Geographical Background of the Jordan Water Dispute." In C. Fisher (ed.), *Essays in Political Geography,* pp. 225–242. London: Methuen.

Brecher, Michael. 1974. *Decisions in Israel's Foreign Policy.* London: Oxford University Press.

Broek, Jan. 1941. "The Problem of 'Natural Frontiers.'" In Committee on International Relations, *Frontiers of the Future,* pp. 3–20. Berkeley: University of California Press.

Bullock, John, and Adel Darwish. 1993. *Water Wars: Coming Conflicts in the Middle East.* London: St. Dedmundsbury Press.

Butler, R., and J. Bury, eds. 1958. *Documents on British Foreign Policy: 1919–1939,* 29 vols. London: Her Majesty's Stationery Office.

Caplan, Neil. 1993. *The Lausanne Conference, 1949: A Case Study in Middle East Peacemaking.* Occasional Paper No. 113. Tel Aviv: Moshe Dayan Center for Middle Eastern and African Studies.

Cohen, Saul. 1986. *The Geopolitics of Israel's Border Question.* Boulder, Colo.: Westview.

Cooley, John. 1984. "The War over Water." *Foreign Policy* 54 (Spring 1984): 3–26.

Davis, Uri, Antonia Maks, and John Richardson. 1980. "Israel's Water Policies." *Journal of Palestine Studies* 9 (2): 3–32.

Dillman, Jeffrey. 1989. "Water Rights in the Occupied Territories." *Journal of Palestine Studies* 19 (1): 46–71.

Donnan, Hastings, and Thomas Wilson. 1994. "An Anthropology of Frontiers." In H. Donnan and T. Wilson (eds.), *Border Approaches: Anthropological Perspectives on Frontiers,* pp. 1–14. Lanham, Md.: University Press of America.

Douglass, J. 1985. "Conflict between States." In M. Pacione (ed.), *Progress in Political Geography,* pp. 77–110. London: Croom Helm.

Esco Foundation. 1947. *Palestine: A Study of Jewish, Arab, and British Policies.* New Haven, Conn.: Yale University Press.

Falah, G., and D. Newman. 1995. "The Spatial Manifestation of Threat: Israelis and Palestinians Seek a 'Good' Border." Presented at the annual meeting of the Association of American Geographers, Chicago, 14–18 March.

Falkenmark, Malin. 1989. "The Massive Water Shortage in Africa—Why Isn't It Being Addressed?" *Ambio* 18 (2): 112–118.

Foucher, Michel. 1987. "Israel/Palestine, Which Borders? A Physical and Human Geography of the West Bank." In P. Girot and E. Kofman (eds.), *International Geopolitical Analysis,* pp. 158–195. London: Croom Helm.

————. 1989. "Cross-Border Interactions: Realities and Representations." In Conference Proceedings: International Boundaries and Boundary Conflict Resolution, Durham, England, 14–17 September.

Fromkin, David. 1989. *A Peace to End All Peace: The Fall of the Ottoman Empire and the Creation of the Modern Middle East.* New York: Avon.

Garbell, Maurice. 1965. "The Jordan Valley Plan." *Scientific American* 212 (3): 23–31.

Garfinkle, Adam. 1994. *War, Water, and Negotiation in the Middle East: The Case of the Palestine-Syria Border, 1916–1923.* Tel Aviv: Tel Aviv University Press.

Gleick, Peter. 1993. "Water and Conflict: Fresh Water Resources and International Security." *International Security* 18 (1): 79–112.

Goldschmidt, M., and M. Jacobs. 1958. *Precipitation over and Replenishment of the Yargon and Nahal Taninim Underground Catchments.* Jerusalem: Hydrologic Service.

Hareven, Alouph, ed. 1977. *Arab Positions Concerning the Frontiers of Israel.* Tel Aviv: Tel Aviv University Shiloah Center for Middle Eastern and African Studies.

Hartshorne, Richard. 1950. "The Functional Approach in Political Geography." *Annals of the Association of American Geographers,* vol. 40, 1950. Reproduced in Roger Kasperson and Julian Minghi (eds.), *The Structure of Political Geography,* pp. 34–49. Chicago: Aldine, 1969.

Hillel, Daniel. 1994. *Rivers of Eden: The Struggle for Water and the Quest for Peace in the Middle East.* Oxford: Oxford University Press.

Hof, Frederic. 1985. *Galilee Divided: The Israel-Lebanon Frontier, 1916–1984.* Boulder, Colo.: Westview.

————. 1995. "The Yarmouk and Jordan Rivers in the Israel-Jordan Peace Treaty." *Middle East Policy* 3 (4): 47–56.

Holdich, Thomas. 1916. "Political Boundaries." *Scottish Geographical Magazine* 32 (1916): 497–507.

Homer-Dixon, Thomas. 1991. "On the Threshold: Environmental Changes as Causes of Acute Conflict." *International Security* 16 (2): 76–116.

————. 1994. "Environmental Scarcities and Violent Conflict." *International Security* 19 (1): 4–40.

Inbar, Moshe, and Jacob Maos. 1984. "Water Resource Planning and Development in the Northern Jordan Valley." *Water International* 9 (1): 18–25.

Ingrams, Doreen. 1972. *Palestine Papers, 1917–1922.* London: Murray.

Institute for Palestine Studies. 1970. *International Documents on Palestine: 1967.* Beirut: IPS.

Isaac, Jad, and Hillel Shuval, eds. 1994. *Water and Peace in the Middle East.* Amsterdam: Helsevier.

Johnson, Douglas. 1917. "The Role of Political Boundaries." *Geographical Review* 4 (3): 208–213.

Jones, Stephen. 1945. *Boundary-Making: A Handbook for Statesmen, Treaty Editors and Boundary Commissioners.* Washington, D.C.: Carnegie Endowment for International Peace.

————. 1954. "A Unified Field Theory of Political Geography." *Annals of the Association of American Geographers,* vol. 64. Reproduced in Roger Kasperson and Julian Minghi (eds.), *The Structure of Political Geography*, pp. 50–56. Chicago: Aldine, 1969.

Kasperson, Roger, and Julian Minghi, eds. 1969. *The Structure of Political Geography*. Chicago: Aldine.

Khativ, Ahmed, and Gamal Khativ. 1988(?). *My Village and the Days*. Self-published.

Kliot, Nurit. 1994. *Water Resources and Conflict in the Middle East*. London: Routledge.

Kolars, John. 1992. "Water Resources of the Middle East." *Canadian Journal of Development Studies* (special issue): 1–24.

Laqueur, Walter. 1967. *The Road to Jerusalem: The Origins of the Arab-Israeli Conflict 1967*. New York: Macmillan.

Libiszewski, Stephan. 1995. *Water Disputes in the Jordan Basin Region and Their Role in the Resolution of the Arab-Israeli Conflict*. Occasional Paper No. 13. Zurich: Center for Security Studies and Conflict Research.

Lonergan, Stephen C., and David B. Brooks. 1994. *Watershed: The Role of Fresh Water in the Israeli-Arab Conflict*. Ottawa: International Development and Research Centre.

Lowdermilk, Walter. 1944. *Palestine: Land of Promise*. New York: Harper and Bros.

Lowi, Miriam. 1993. *Water and Power: The Politics of a Scarce Resource in the Jordan River Basin*. Cambridge: Cambridge University Press.

MacBride, Sean. 1983. "Israel in Lebanon: The Report of the International Commission to Enquire into Reported Violations of International Law by Israel during Its Invasion of the Lebanon." n.p.: n.p.

McCarthy, Justin. 1990. *The Population of Palestine: Population Statistics of the Late Ottoman Period and the Mandate*. New York: Columbia University Press.

Main, Chas. T., Inc. 1953. *The Unified Development of the Water Resources of the Jordan Valley Region*. Knoxville: Tennessee Valley Authority.

Medzini, Meron, ed. 1976. *Israel's Foreign Relations*. Jerusalem: Ministry for Foreign Affairs.

Meinertzhagen, Richard. 1959. *Middle East Diary: 1917–1956*. New York: Yoseloff.

Mekorot Water Company, Ltd. 1944. *The Water Resources of Palestine*. Tel Aviv: Mekorot.

Minghi, Julian. 1963. "Boundary Studies in Political Geography." *Annals of the Association of American Geographers*, vol. 53, 1963. Reproduced in Roger Kasperson and Julian Minghi (eds.), *The Structure of Political Geography*, pp. 140–160. Chicago: Aldine, 1969.

Murakami, Masahiro. 1995. *Managing Water for Peace in the Middle East: Alternative Strategies*. Tokyo: United Nations University Press.

Myers, N. 1993. *Ultimate Security: The Environmental Basis of Political Stability*. New York: W. W. Norton and Company.

Naff, Thomas. 1992. *Water Scarcity, Resource Management, and Conflict in the Middle East*. American Association for the Advancement of Science Report 92-33S. Washington, D.C.: Committee on Science and International Security, American Association for the Advancement of Science.

————, and Ruth Matson, eds. 1984. *Water in the Middle East: Conflict or Cooperation?* Boulder, Colo.: Westview Press.

Neff, Donald. 1994. "Israel-Syria: Conflict at the Jordan River, 1949–1967." *Journal of Palestine Studies* 23 (4): 26–40.

O'Loughlin, John, ed. 1994. *Dictionary of Geopolitics.* Westport, Conn.: Greenwood.

Pedhatzor, Reuven. 1989. "A Shared Tap." *Ha'aretz,* May 3.

Prescott, J. 1965. *The Geography of Frontiers and Boundaries.* Chicago: Aldine.

————. 1987. *Political Frontiers and Boundaries.* London: Allen and Unwin.

Ra'anan, Uri. 1955. *The Frontiers of a Nation: A Re-examination of the Forces Which Created the Palestine Mandate and Determined Its Territorial Shape.* Westport, Conn.: Hyperion Press.

Sachar, Howard. 1969. *The Emergence of the Middle East: 1914–1924.* New York: Knopf.

————. 1979 (vol. 1) and 1987a (vol. 2). *A History of Israel.* New York: Knopf.

————, ed. 1987b. *The Rise of Israel,* 39 vols. A facsimile series reproducing 1,900 documents. New York: Garland.

Schiff, Ze'ev. 1995. "They Are Forgetting the Golan's Water." *Ha'aretz,* June 7.

Schmida, Leslie. 1983. *Keys to Control: Israel's Pursuit of Arab Water Resources.* Washington, D.C.: American Educational Trust.

Schwartz, Yehoshua, and Aharon Zohar. 1991. *Water in the Middle East: Solutions to Water Problems in the Context of Arrangements between Israel and the Arabs.* Tel Aviv: Jaffee Center for Strategic Studies.

Shalev, Aryeh. 1994. *Israel and Syria: Peace and Security on the Golan Heights.* Jaffee Center for Strategic Studies Study No. 24. Jerusalem: Jerusalem Post Publishing.

Shemesh, Moshe. 1988. *The Palestinian Entity 1959–1974: Arab Politics and the PLO.* London: Frank Cass.

Shuval, H. 1992. "Approaches to Resolving the Water Conflicts between Israel and Her Neighbors: A Regional Water-for-Peace Plan." *Water International* 17 (3): 133–143.

Slater, Robert. 1991. *Warrior Statesman: The Life of Moshe Dayan.* New York: St. Martin's Press.

Soffer, Arnon. 1994. "The Litani River: Fact and Fiction." *Middle Eastern Studies* 30 (4): 963–974.

Starr, Joyce. 1991. "Water Wars." *Foreign Policy* 82 (Spring 1991): 17–36.

Stauffer, Thomas. 1982. "The Price of Peace: The Spoils of War." *American-Arab Affairs* 1 (Summer 1982): 43–54.

Stein, Janice, and Raymond Tanter. 1980. *Rational Decision-Making: Israel's Security Choices, 1967.* Columbus: Ohio State University Press.

Stevens, Georgiana. 1965. *Jordan River Partition.* Stanford, Calif.: Hoover Institution.

Stork, Joe. 1983. "Water and Israel's Occupation Strategy." *MERIP Reports* 116 (13): 19–24.

Tamir, Avraham. 1988. *A Soldier in Search of Peace: An Inside Look at Israel's Strategy.* London: Weidenfeld and Nicolson.

Tarnopolsky, N. 1996. "Water Damps Hopes for Deal with Syrians." *Forward,* 5 January.

United States House of Representatives. 1990. "Hearing on Middle East Water Issues in the 1990's." June.

Waterman, Stanley. 1984. "Partition: A Problem in Political Geography." In P. Taylor and J. House (eds.), *Political Geography: Recent Advances and Future Directions,* pp. 98–116. London: Croom Helm.

Weinberger, Gabriel. 1991. *The Hydrology of the Yarkon-Taninim Aquifer.* Jerusalem: Israel Hydrologic Service.

Weizmann, Chaim. 1968. *The Letters and Papers of Chaim Weizmann,* ed. Leonard Stein. London: Oxford University Press.

Wishart, David. 1989. "An Economic Approach to Understanding Jordan Valley Water Disputes." *Middle East Review* 21 (4): 45–53.

Wolf, Aaron T. 1995a. *Hydropolitics along the Jordan River: Scarce Water and Its Impact on the Arab-Israeli Conflict.* Tokyo: United Nations University Press.

———. 1995b. "International Water Dispute Resolution: The Middle East Multilateral Working Group on Water Resources." *Water International* 20 (3): 141–150.

———. 1996. *Middle East Water Conflicts and Directions for Conflict Resolution.* 2020 Vision Initiative Monograph No. 12. Washington, D.C.: International Food Policy Research Institute.

World Press Review. 1995. "Next, Wars over Water?" November, pp. 8–13.

5. A Popular Theory of Water Diversion from Lebanon
TOWARD PUBLIC PARTICIPATION FOR PEACE

Hussein A. Amery

The enduring war to be waged in the Middle East is one of images, of messages, of what people in the street believe is happening. —J. AbiNader

INTRODUCTION

Allegations of environmental or political coverups are neither new nor rare. A relevant current example pits the water-rich state of Canada against the thirsty southwestern areas of the United States. Some Canadians fear that the North American Free Trade Agreement (NAFTA) treats water like any other tradable commodity, which will then force their country to export water resources to the signatories of the agreement (*Globe and Mail* 1993). Another example involves reports about Israeli technical assistance to Ethiopia to help it build dams on the Nile River. Although these reports were not substantiated, they nevertheless provoked much anger and suspicion in Egypt. There is a list of disputed data in the Jordan River basin area: the exact amount of water consumed by Israelis and Palestinians (*Economist* 1995, 41), the groundwater recharge that Gaza receives (Arar 1993), and the flow of the Litani River (Soffer 1993). Finally, an Israeli scientist recently corrected his earlier work by stating that climatic changes were "responsible for the destruction of the agricultural economy and the invasion of the sand dunes along the coastal plain of Israel" during the eighth century. He had erroneously deduced, about one decade after Israel was created, that the culprits were the Arabs who assaulted the area during the seventh and eighth centuries (Issar 1995, 350–351). After allowing for advances in technology and scientific learning that permit one to investigate matters with greater clarity and depth, it is also likely that the earlier conclusions were influenced by the perception at that time that Arabs are fierce and barbaric. Power and cultural differentials, spectacular victories and defeats, governments with little or no legitimacy, spatial restrictions on citizens' movement, and historical animosities are conducive to various types of (mis)perceptions, including environmental allegations between two or more population groups. This may in turn color the approach and findings of some "scientific" studies. The degree of truth in rumors becomes immaterial. Consequently, adverse perceptions linger on among the people and some of their political leaders, which could affect the socioeconomic and political interactions between states.

The Hasbani and Wazzani Rivers spring from Lebanon, thus making Lebanon a riparian on the Jordan River. The entire flow and catchment area of the Litani River are, on the other hand, inside Lebanon, thus making it an entirely Lebanese river. A large number of analysts from a variety of disciplines have for decades been arguing that Israel is getting ready to divert, is already diverting, or covets the waters of Lebanon's Litani River—parts of which have been under Israeli occupation since 1978 (Figure 5.1). That is to say, they believe in the existence and feasibility of a Litani-to-Israel diversion scheme, or the desire to establish one. On the other hand, another group of commentators rejects allegations of a unilateral water diversion: instead, some of them propose a negotiated water "sharing" or purchasing arrangement with Lebanon. This rejection, as well as the breadth and persistence of the diversion explanations, especially among the people, justifies the use of the word "theory" in this chapter's title. According to R. Hessler (1992, 10), a theory is "another word for explanation. Research begins with a question, and theory is the attempt to answer the question or to explain the research problem." The *Oxford Paperback Dictionary* states that a theory is "a set of ideas formulated (by reasoning from known facts) to explain something" or "an opinion or supposition." The issue in this chapter is more about what is perceived to be by the people than about what is.

Why does this popular theory of alleged illicit water appropriation linger unresolved to the satisfaction of skeptics in Lebanon and beyond? How explicit and precise is the corroborating evidence? In addressing these questions, this chapter offers a critical evaluation of published works on the theory of water diversion from the Litani River by drawing on relevant Arabic- and English-language literature. The latter body of research represents work done by Western and Israeli scholars who, surprisingly, have never considered the opinions of local residents, water practitioners, academics, or politicians in Lebanon or the Arab world. This chapter represents the first attempt at understanding this popular theory by using such Arabic sources from published and oral accounts. These data are used to examine the accuracy of the evidence used to

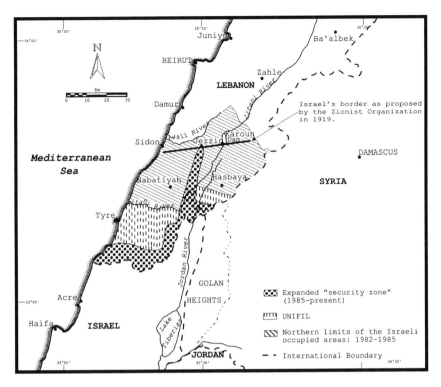

FIGURE 5.1. The Litani River and the Israeli-occupied "Security Zone."

corroborate the theory, and to reveal the level of public knowledge of it inside Lebanon.

The chapter's main findings are that the meager *evidence* against the theory is less than convincing to a justifiably skeptical public in Lebanon, and especially to those living in the South. On the other hand, the ample evidence in support of this popular theory, while strong, is sometimes contradictory and hence inconclusive. Moreover, the water diversion theory, irrespective of its validity, is shown here to be deeply ingrained in the minds of many in Lebanon. Given that a peace agreement between Lebanon and Israel will eventually be reached, reconciliation and "warm peace" (as opposed to the Israel-Egypt model of "cold peace") would become a high priority. Consequently, this chapter proposes the process of public participation as a potential avenue for dealing with this popular theory. The strategies that a public inquiry may choose to take vary from (1) indirectly dealing with the theory through water resources management in southern Lebanon to (2) directly confronting it through an investigation of the theory itself. (It is worth noting that diverting waters from the Litani and other southern rivers for use in the city of Beirut, the West Bank, Jordan, and some oil-rich Gulf states is an idea that has been in circulation for a few years.) The philosophy of public participation revolves around involving the public in the development of decisions that affect them (FEARO 1994). Participation has educative and expressive dimensions as well. It helps to develop reasonableness and sensitivity in people by broadening their outlook through firsthand experience and information. Public inputs highlight impacts that may not have been considered, and may provide innovative and viable (impact-mitigating) ideas which if implemented would result in a water management strategy that is supported by the majority. This inclusive approach to resource planning is believed to assist in dealing with the widely held diversion theory, which, if unchecked, will continue to be a source of distrust between the Israelis and Lebanese.

A CONCEPTUAL UNDERSTANDING OF THE ISSUE[1]

People's perception of the security of their natural resources is influenced by the design of the built environment. In areas where turmoil and political instability are present, people regularly evaluate their risk of victimization by constant monitoring and scanning of their immediate environment for cues of danger. These cues from their surroundings are evaluated, and untoward signs of alarm are responded to (Goffman 1971). People do not recognize all events as alarming, and respond only to those perceived as unsafe or threatening (Nasar and Fisher 1992).

The time, location, and spatial scope of an event influence the spatial distribution of signals of alarm. Thus, some situations are perceived as being alarming while others are not (Nasar and Fisher 1992). This may partly explain the surge in reported attempts of water diversion in 1984 (details below), two years after Israel had occupied the largest area of the Litani watershed to that date. The 1982 war resulted in the substantial expansion of the security zone which was created in 1978. Israel's occupation of about 40 percent of the country (1982–1984) aggravated the perception of threat because (1) this expanded spatial control created a sense of boundedness which physically and psychologically cut off a larger segment of the population from other areas. More importantly, (2) the added territory permitted the occupying "enemy" to operate from positions of prospect and refuge.[2] These mean that the suspected party was able to access sufficient geographical and strategic areas to affect or carry out a scheme (prospect) with low probability of being exposed (refuge) to the public. Like rugged terrain, space that is regulated or designed by a distrusted actor to be off-limits hinders people's ability to see what might be lurking there. The public's concern is based on the fact that the physical environment provides the suspected party with refuge. This in turn creates uncertainty about what could be occurring. Consequently, the geographical expansion into neighboring territory, and the capacity for "out-of-sight" activity, trigger signals of potential danger for the subjugated population because of its artificially ("militarily") limited ability to affect the situation. After the Negev nuclear reactor leaked radioactive wastewater into a nature reserve, the Israeli government's assurance did not satisfy a skeptical environmentalist. She said: "From the history of nuclear events around the world, we know that absolute secrecy almost automatically guarantees irresponsibility and total disregard of the rights of citizens" (Collins 1993, 24).

Signs of alarm regarding the Litani water resources are expressed by Lebanese and Arabs, and by high-ranking officials in Lebanon (and the region), through many media and educational channels. An alarmist message reaches many Lebanese due to the high level of literacy, and the vibrant press in the country. These helped in popularizing the theory and affixing it in the minds of citizens. Najah Wakeem, a Lebanese MP, Raymond Edih, a prominent veteran politician, and other notables of varying political and religious stripes in Lebanon believe that Israel "covets" or is "diverting" water from southern Lebanon. Moreover, the Secretary General of the Arab League, Ismat Abdul Majeed, acknowledges the "threat" to Lebanon's waters (Nabeel 1992; see also Haikal 1969 and *an-Nahar* 1994, 5). The roots of this concern go back to the early years of this century.

HISTORICAL SETTING

Since 1916, Zionist plans called for the Litani River to form at least the northern boundary of their future state (Naber 1968; Weisgal 1977). The desired and demanded borders of the envisioned state were often delineated more by its perceived hydrological "needs" than by security concerns. The lower reaches of the Litani River, the Hasbani and Wazzani Rivers, were invariably included as being vital to the new state (Hof 1985; Amery and Kubursi 1994). The new state of Israel failed to achieve all of its territorial demands from the British mandatory authority. The Arab-Israeli war that ensued after Israel was formally declared in 1948 resulted in, among other things, the occupation of Lebanese territory reaching the western bend of the Litani River (Amery 1998). The 1949 armistice agreement between Lebanon and Israel ended that war and resulted in a mutually agreeable demarcation of their international border. Despite this armistice agreement, Israeli government officials in the 1950s discussed at some length a geopolitical and military strategy for regaining the vacated border-to-Litani lands. The then–Prime Minister of Israel, Moshe Sharett, quotes Moshe Dayan (Israel's Chief of Staff and later Defense Minister) as having said in 1954 that the Israeli army would enter Lebanon and occupy and annex to Israel the territory south of the Litani River (Rabinovich 1985, 163; for more details see Eisenberg 1994 and Assi 1968). Perceptions of Israel's covetous interests in Lebanon's water resources ebbed and flowed through time but never disappeared, certainly not from the minds of the Lebanese population.

After Israel was created, Soffer (1993) correctly states, only once did it try to include the water flow of the Litani River in the Jordan River system; this was during the Eric Johnston negotiations (1953–1955) regarding the reallocation of the Jordan waters. "Since then, Israel has not demanded officially [sic] use of the Litani, and has not proposed using the river in any international forum" (Soffer 1993, 968). For the sake of completeness, it must be noted that Israel's Prime Minister Sharett said during the third round of the Johnston talks that "while Israel had accepted the *temporary* exclusion of the Litani," it was against any water developments on the (lower) Litani, so as not to "preclude the eventual joining of its waters with other rivers of the Jordan system" (Lowi 1993, 239n. 69; my emphasis). Consequently, Lebanon's inability to develop its southern region is seen as being related to (1) Israel's continued covetous interest in the waters of the Litani, and (2) its believed objection, at the international stage, to the river's utilization. With respect to the former, the Lebanese President Bishara al-Khoury stated that "The American policy in the area does not allow the irrigation of the South without tacit cooperation from Israel and its sharing the water of the [Litani] river" (Sharif 1978, 21). Regarding the latter, when the National Authority for the Litani River was denied funding from the International Bank for Reconstruction and Development for the purpose of using the Litani waters to irrigate the South, Lebanon accused Israel of objecting to and effectively vetoing the project (Baalbaki 1985). The predominantly Shia Muslim residents of southern Lebanon have long felt neglected and marginalized by the central government, which continues, although to a lesser extent now, to be dominated by the Maronite Christians and Sunni Muslims. Moreover, Lebanon's "inability" to develop the South was aggravated by the Palestinian-Israeli low-intensity conflict, and later by the Lebanese civil war. Lonergan (1993, 213) argues that Israel had been interested in reaching an agreement with the Lebanese government, so as to import some of the waters of the Litani River. However, he adds, given its current water stress, "Israel may seriously consider other options to divert the Litani regardless of the legalities or international repercussions." Given the state of war between the two countries, the massive difference in military and political power, and the Israeli occupation of the water-rich area in question, an "official" request for an emotionally charged resource is not the likely route that would be taken.

A book helped in rekindling the theory in Lebanon one year after Israel occupied the West Bank and the Golan Heights in the 1967 war. The seemingly prophetic book, published in Lebanon ten years prior to Israel's occupation and creation of the "security belt," maintained that Israel intended to occupy southern Lebanon so as to control its water resources (Assi 1968). At another level, Muhammed Hassanain Haikal, a distinguished Egyptian journalist, commentator, and confidant of the late Gamal Abdel Nasser, reported in a Lebanese newspaper the following in 1969: the Egyptian president had told Sabri Hamadeh, Lebanon's Speaker of the Parliament (Chamber of Deputies), that "Israel is after the springs of the Wazzani and Litani rivers, and consequently, Lebanon is in the crux of the [Arab-Israeli] conflict. We hope that everyone in Lebanon pays attention to this danger, and makes the necessary preparations to avert this threat" (Haikal 1969, 3). This shows that the theory was being discussed in the media and by academics, and was accepted by high-ranking officials shortly after the 1967 war, in which Lebanon was not involved. Consequently, Israel's 1978 invasion of Lebanon and its occupation of the "security zone" was understood by many as closely corresponding with what Israeli strategists had discussed just over twenty years earlier (Amery and Kubursi 1994).

The occupied zone inside southern Lebanon was officially described as having the objective of providing security for Israel's northern residents from Palestinian guerrillas. On the other hand, this invasion, which was not perceived by most Lebanese and Arabs as being related to Israel's physical security, reinforced and transformed perceptions of the Lebanese. Israel's historic interests in diverting the waters of the Litani were seen as having moved a step closer to being realized (Agnew and Anderson 1992). The majority of the people in Lebanon still perceive this occupation as yet another manifestation of Israeli expansion-ism. This view corresponds with a similar long-standing Arab percep-tion. On the other hand, the majority of Lebanese view the occupation as a functional maneuver that gave Israel the opportunity and the lands necessary to set in motion its long-awaited diversion plan. This was so prominent that within days of the Israeli invasion the Lebanese press, and even international newspapers such as *The Christian Science Moni-tor,* reported the assumed undeclared hydrological dimension of the war and subsequent occupation of the security zone (Cooley 1978). Clearly, Israel's prospect was greatly enhanced by the occupied zone.

The theory gained greater affirmation and relevance after Israel's 1982 invasion of Lebanon, which was more encompassing geographi-cally and temporally. Shortly after Israel occupied about 40 percent of Lebanon, an Israeli government official confirmed that its army engineers had carried out "seismic soundings and surveys . . . to assess the feasibil-ity of a Litani diversion tunnel" (Stork 1983, 24; see also Cooley 1984 and Stauffer 1985). Lonergan (1993, 214), like many others, argues that "There is a great deal of support for the theory that water had a key role in the 1982 invasion of Lebanon by Israel." Similarly, Nasrallah (1990, 17) states that "One of the multiple objectives of Israel's invasion of Leb-anon in 1982 was to control the waters of the Litani." Shortly after this invasion, the former American ambassador to Saudi Arabia during the 1970s, James Akins (1982, 36), wrote that "the diversion of the Litani River into Israel is almost a certainty." The chairman of the National Authority for the Litani River said that in June of 1982 Israel had seized "all the hydrographic data on the [Qir'awn] dam and the river," a fact acknowledged by Israel, which considers them "legitimate items of mili-tary intelligence" (Stork 1983, 24; see also Naff 1993 and Cooley 1984). A 1983 report by the International Law Commission states that "Israel harbored certain territorial aspirations toward Lebanon, the most sig-nificant of which may have been its intention to obtain access to or pos-sibly control over the waters of the Litani River" (ESCWA 1991, 3). Is-rael will not withdraw from the occupied zone, according to the General Director of Israel's Foreign Ministry, until it receives a share of the waters of southern Lebanon (this was in a letter he had sent to the American mediator, Morris Draper, quoted in Khalifeh, n.d.). The essence of this is consistent with Naff (1993). Finally, P. Beaumont (1994, 17) states that "The Israeli government continues to deny that it is now using the waters of the Litani, but various reports from people in the area suggest that this may not be the case."

Soffer (1993) and Wolf (1995) reject the existence of *any* hydrological intention behind Israel's territorial expansions. They see Israel's physical security as a primary motivation for Israel's wars. While many agree with their no-diversion argument, the unequivocal tone is almost unique to Wolf and Soffer. For example, Schofield (1993, 153), like many other such writers, argues that Israel does not "appear to be extracting water [from the Litani] at the present," but its impending higher water demands will make the Litani a "key element in Israeli water security planning" (a similar view is held by Naff 1993). Similarly, Hof (1985) and A. Norton argue that Israel's occupation of southern Lebanon transcends the security of its northern settlements. Norton adds that "common sense argues against assuming chaste motives" (quoted in Schofield 1993, 160).

Israel altered the geopolitical environment by occupying in 1978 a security zone in order to, as it declared, protectively cut off its citizens from Palestinian guerrilla attacks. This geographical buffering of one population left another feeling vulnerable and deeply suspicious because it was cut off from its physical, social, and political environment.

Was water an objective in Israel's occupation of southern Lebanon? A plausible answer that is conceptually grounded emerges from the work of Horowitz (1983). He states that predetermined objectives are quite different from ancillary ones. There are "objectives for which one does not initiate war, but are rather appended to the original objective of a defensive or deterrent war." This is based on the idea of open objectives or "exploitation" of success, a concept accepted by the Israeli military establishment (Horowitz 1983, 92). An Israeli magazine quotes government and military people in Israel as saying that "former Prime Minister Menachem Begin had water in mind as a consideration while planning the 1982 invasion of Lebanon" (Ben Shaul 1990, 6). Others go further by asserting that the Israeli cabinet, and the country's water planning agency, Tahal, had discussed diverting the waters of the Litani and other rivers during the 1982 invasion (Naff 1993; see also Schofield 1993). Procuring water resources from another state might not in itself have been worth initiating a full-scale war over, but it was an ancillary goal of Israel's invasions of Lebanon. Occupying southern Lebanon would give Israel stronger leverage with respect to the Litani, and diverting even small amounts of water from the river would internationalize it. One or both of these factors would facilitate the river's integration into a water-allocation scheme involving the water resources of the entire Jordan River basin.

Arguments presented in this section are based on broad assertions that the objective of Israel's occupation of southern Lebanon is a desire to acquire greater water resources. These do not shed light on the question: Is any water actually being diverted southward to Israel?

WATER DIVERSION THEORY: FACT OR FABRICATION?

Determining whether Israel has hydrological intentions in south Lebanon is essentially an elusive issue; the existence of water diversion structures should not be. Evidence for and against the water diversion theory is examined in the following section.

The diversion of water from southern Lebanon to Israel is seen as an unfounded allegation by Soffer (1993), Wolf (1995), Naff (1993), and others. Gathering and presenting evidence regarding the physical reality on the ground, therefore, becomes crucial.

There have been numerous testimonies and supportive evidence that lend credence to the theory. Some evidence is, however, understandably sketchy and not sufficiently quenching because the suspected sites for a diversion are under occupation and therefore made inaccessible to local residents and outsiders. Statements by officials and eyewitness accounts represent some of the reported evidence. For instance, in a speech designed to convince Lebanon's Parliament to adopt the 17th of May (1993) peace agreement that Lebanon and Israel had tentatively agreed to, Lebanon's Foreign Minister Eli Salem stated in the Parliament that "Israel had completed all of the necessary plans and scientific studies for a water diversion from Litani" and hinted that these plans would be implemented if the deputies voted against the agreement—according to which Israeli troops would leave Lebanon (E'taani 1984, 220). The agreement was not ratified. One year later, the Speaker of Lebanon's Parliament, Munir Abu Fadil, stated that "Israel had completed building a large tunnel and placing pumping stations, and may have started preliminary testing of water diversion" from the Litani River (al-Fursan 1984, 6). Similar facts were echoed by two leading magazines which specified the tunnel as connecting Tell al-Nahas with al-Khardali and Lake Tiberias (al-Tadamun 1984; al-Hawadith 1984). Bulloch and Darwish (1993, 42) quote a UN report that concurs with the theory. More recently, the Lebanese MP Najah Wakeem presented a picture that was published on the front page of a leading independent Lebanese newspaper a decade earlier (an-Nahar, 4 August 1984). It shows Israeli tunneling operations to divert waters from the Litani (*an-Nahar* 1994, 5). There were also reports of "excavation works [by the Israeli army] near the Litani," and "non-seasonal falls in the water level" in the Kassimiyeh River downstream (Abboud 1986, 4). Bulloch and Darwish (1993, 41–42) report that "local farmers say that Israel has built a tunnel from the Litani to the Hasbani, thus diverting water into the Jordan." In a book for the Department of the Army in the United States, T. Collelo (1989, 117) states that "In the late 1970s and early 1980s, Lebanese officials reported that small tributaries of the Hasbani River were being diverted to Israel near the northern town of Metulla. Independent water analysts stated that after the 1982 invasion, Israel engaged in a much more serious diversion of Lebanese waters by attaching stopcocks at a pumping station on the Litani River. The stopcocks were designed to switch at least part of the flow—which is generated entirely within Lebanon—to Israel via a specially constructed pipeline." The "actual" water diversion route is even mapped out (Figure 5.2) and published in academic sources and in the mainstream media in Lebanon.

A recent report by the United Nations' Economic and Social Commission for Western Asia (ESCWA) states that in 1978 Israel started to pump 150 million cubic meters (mcm) of water from the Litani annually, and after the 1982 invasion it "drilled an 18-km tunnel which links the Litani to Israel." It also states that Israel alone has been using the waters of the Wazzani River, which amount to 65 mcm per annum (ESCWA 1994, 8). The American academic Thomas Naff (1990, 7) testified to the House of Representatives of the United States that "Israel is presently conducting a large-scale operation of trucking water to Israel from the Litani River." The director of the Lebanon-based Byblos Centre for Research, Nabeel Khaleefeh, states that Israel had "recently placed rocks in the [channel of the] Litani in such a way so that when its water level rises, water flows into the [already built] tunnel without a noticeable change in the [downstream—Kassimiyeh] flow of the river" (A'yash 1995, 5). This is consistent with Cooley (1984, 22–23), a correspondent for the American Broadcasting Company (ABC News), who quotes an American military observer as saying that the Israeli army was burying pipes to secretly siphon water into Israel "without affecting the measured flow of the Litani." Finally, a large map in an Israeli newspaper carried the following caption inside a box with an arrow pointing to the Litani River: "Israelis pumping water from occupied southern Lebanon to Israel" (Davis 1990, 9). This text is neither explained nor reiterated in the accompanying news story. The preceding sample of observations

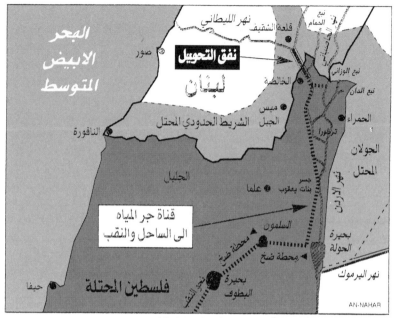

البحر الابيض المتوسط

نهر الليطاني   قلعة الشقيف   نبع النمام

نفق التحويل

صور

الخالصة   لبنان   نبع الوزاني   نبع الدان

ميس الجبل   الشريط الحدودي المحتل   الحمراء   ترشيشا

الناقورة

الجولان المحتل   نهر الاردن

الجليل   جسر بنات يعقوب   علما

قناة جر المياه الى الساحل والنقب

السلمون   بحيرة الحولة

محطة طبريا   نهر اليرموك

محطة ضخ

حيفا   فلسطين المحتلة   بحيرة النقطة   بحيرة البطوف

AN-NAHAR

خريطة للمهندس حسين رمال أظهرت انفاقاً في الاراضي اللبنانية لنر المياه الى اسرائيل.

FIGURE 5.2. A map from a respected mainstream Lebanese newspaper showing the location of the diversion tunnel from the Litani. (The white-on-black Arabic caption reads: "diversion tunnel." *an-Nahar,* 21 February 1995, p. 5.)

amounts to corroborative Arab, American, and Israeli evidence of the water diversion postulate. However, there are lingering questions and issues which explain why many researchers continue to reject the theory.

A detractor of the theory, A. Soffer (1993), has strongly and emotionally rebuked those who even hint that Israel may be diverting any water from the Litani River. "Israeli invasions of southern Lebanon [1978, 1982] were a direct result of PLO attacks on Israel, and in no way does Israel control the Litani zone [sic], nor has it ever transferred even one cup of water from the Lebanon to Israel" (Soffer 1993, 968). The area of the Litani that came under Israeli control in 1978 was in fact augmented after the 1982 war. Moreover, Soffer does not offer any conclusive evidence to the contrary except for two important quotations of Lebanese officials who discussed progress on a major irrigation works on the lower reaches of the Litani, the Kassimiyeh River. Such water development projects are argued to be possible only if there were "sufficient" amounts of water in that portion of the river. This, however, does not preclude diverting relatively small amounts when water discharge is at its peak. Similarly, Kolars (1993, 39) interviewed two UN peacekeeping soldiers and A. Soffer, an Israeli geographer and former high-ranking commander in the Israeli army who had served in Lebanon during the 1982 invasion, and concluded that "there is no convincing evidence that such a [water diversion] tunnel exists." No supporting details whatsoever are provided. Kolars, however, states that "In the case of either trucks or small pipelines the amount of water transferred is physically insignificant, but the action could, nevertheless, be symbolically sensitive" (Kolars 1993, 40).

As evidence to refute the popular theory of water diversion from Lebanon, Wolf (1995) uses his field investigations, a review of satellite photos, air photographs of the Israeli air force, and maps of the Israeli water authority. He, however, merely lists (with no references because, as he told me, of the sensitive sources that he relied upon) the method on which he bases his argument. That is to say, he fails to provide any information about the rigor or extensiveness of the method of investigation used in analyzing the maps and photographs, and whether field inquiries were carried out in order to confirm the findings in the data. For example, were there field studies of specific "highly suspect" water diversion sites? If so, which sites and when were they studied? Moreover, it is puzzling that Wolf does not mention *any* "gray areas" in the analyzed photos or maps. This is a significant omission especially when one considers the Israeli army's extensive engineering and construction activities, which it has been carrying out in the security zone since its occupation of the security zone in 1978.[†]

Other pillars of Wolf's (1995) no-diversion argument are (1) reviews of Israeli hydrological records (again, no details are given) and (2) interviews with an Israeli army liaison officer and with a senior Israeli military strategist who was involved in planning the 1967 and 1982 wars. The officers "insist that water was not, even incidentally, a factor in these two wars" (76). One officer quotes Sa'ad Haddad as boasting about quashing "Israel's plans in 1979 to divert the Wazzani springs" (75). This unexplained comment (one short year after the occupation of southern Lebanon) contradicts Wolf's assertion that those hydrological considerations did not in any way contribute to the occupation of southern Lebanon. The quoted Israeli officers were shown elsewhere (Amery forthcoming 1999) to have provided Wolf with inaccurate data which called into question the validity of at least some of the information obtained from military or ex-military sources. This is explained by what a planner of many of Israel's water projects had said: "Any actions to divert water from Lebanon into Israel would be under military control and, as such, a military secret" (Stork 1983, 24). Another example of military secrecy that aggravates diversion suspicions originates in Lebanon. Lebanon's Defense Minister Adel U'sayran requested and received from Simon Kassiss, the head of military intelligence, a report regarding his agency's assessment of whether or not Israel had diverted the Litani waters. The contents of the decade-old report, which was presented to the council of ministers, remain classified. It is, therefore, normal and understandable for military officers not to reveal the full nature of their activities. Consequently, academics ought to be cautious if and when they rely on such sources, and to scrutinize carefully the gathered information.

In addition to the evidence and intrigue which have raised the profile of the theory and are giving it greater acceptability among the Lebanese, Israeli scientists have repeatedly outlined ways in which the Litani (and other) waters may be used outside the country. One study suggests regional water "integration" which would entail siphoning the surplus coastal and other waters for thirsty states and regions in the Middle East and Israel (Issar 1993, 1). Another study outlines and maps out two routes through which the Litani waters may be diverted (Ben-Shahar et al. 1989, 105). The diversion map, which has been translated into Arabic and reproduced in numerous books and articles in Lebanon, is being cast as evidence of Israel's water designs on Lebanon. Elisha Kally (1991; 1991–1992), a strong proponent of future water transfer, discusses the benefits for Lebanon: financial returns from the sale of water resources, and a greater supply of hydroelectricity from Israel. These scenarios are proposed in the context of a Middle East at peace.

On a regional scale, Lebanon is viewed as having surplus water resources which could be shared. One Israeli wrote that many water planners in his country "look northward" to Lebanon to solve their country's water scarcity (Ben Shaul 1990, 6). Another states that "Relief [from water stress] may come if Israel can strike a deal with Lebanon over water from the Litani, and with Jordan over the largely untapped Yarmouk River" (Davis 1990, 9; see also Schwarz 1990). A Likud minister in the Israeli government stated that Israel would be "glad to buy this little water and make good use of it in northern Galilee" (Cooley 1984, 24). Obliquely reminiscent of the historical claim that the Litani waters are vital for Israel, and can be put to better use rather than being "wasted" into the sea (Hof 1985), an Israeli writes that "The amount of water that daily pours into the sea from the Litani River . . . is equal to Israel's total water needs" (Ben Shaul 1990, 6). These arguments imply that the flow in the lower Litani is sufficient to make a diversion feasible, and support the belief that south Lebanon's water is coveted by Israel.

REACTIONS TO DIVERSION THEORIES

The reactions of Lebanon to this historically persistent water diversion theory have been consistently low-key and oblique, except for once raising the issue before the United Nations. This may be explained partially by the fact that the full extent of Israel's intentions regarding the Litani were not revealed until decades later (see Eisenberg 1994; Hof 1985; Rabinovich 1985). The creation of the National Authority for the Litani River in 1954, with a mandate to develop the waters of the river, amounts to Lebanon's first major unofficial "reaction" to claims prior and subsequent to Israel's creation.

Perhaps the most significant official public reaction came two years after Israel's 1982 invasion of Lebanon. In the summer of 1984, Lebanon officially raised the issue of Israeli water diversion at the Security Council. On the 6th of September, Rachid Fakhoury, Lebanon's ambassador to the UN, stated that his country's "suspicions and fears about Israel undertaking water-diverting activities had been confirmed" (*UN Chronicle* 1984, 12). The inaction which followed was due to the following: (1) a proposed resolution about the matter was vetoed by the United States, (2) Lebanon was in the middle of a civil war, and (3) 40 percent or so of its territory, including a large part of the Litani River area, was occupied by the suspected party. Another international protest occurred in 1990, when Lebanon sent a letter dated 14 April to the United Nations Security Council expressing its concern over Israel's covetous interests in the waters of the Litani (Az-Zu'be 1992).

In response to the above-mentioned 1994 report by the ESCWA, Lebanon's certainty and aggressiveness at the UN in 1984 were replaced with skepticism and reluctance a decade later. The country's Foreign Minister, Faris Buwayz, said that

> Israel's intentions and ambitions in Lebanese lands and water are neither surprising nor new. Information circulated over the past few years is not new. We repeatedly wrote to the United Nations with requests to send fact-finding committees to verify information that we received from several sources. The UN reply clearly emphasized that

there was no evidence of water theft, either on the surface or underground, through pipes, pumps, or tunnels. (*National News*)

He added that as a result of the new ESCWA findings, the Lebanese government requested a UN "experts committee to inspect the al-Khardali area and tell us whether Israel is, or is not, stealing water directly or indirectly. Hence we are waiting for the UN findings" (Beirut Radio, 2 June 1994). Officially, then, the government acknowledges the threat in the form of an Israeli interest in its water resources, dismisses a substantiation of it by a UN agency (ESCWA), and at the same time asks for more research by the UN to substantiate whether a diversion actually exists. It is worth noting that the Speaker of the Parliament, Mr. Nabih Birri, and other parliamentarians have voiced their support for the ESCWA report. Lebanon's muted and divided reaction is likely related to the country's "close ties" to Syria. The Prime Minister of Lebanon, Rafiq Hariri, said that "our relationship with the Syrians is based on two unchanging features: cooperation in internal security matters— and I think most Lebanese see it as effective—and coordinated foreign policies" (*Le Quotidien de Paris* 1994, 11). Consequently, the introduction of politically detracting issues such as illicit water diversion at a time when Syria's peace talks with Israel were going through a delicate stage would have been unwelcome. This reasoning was also used to explain Hariri's "forced" retractions of his resignation as Prime Minister in 1995.

The ESCWA (1994) report, which confirms the existence of water diversion from Lebanon, was so controversial that the day after it was issued to the media, the UN agency released a statement reporting that its findings were based on "published theoretical studies and research, because of the inability to conduct a field survey on location due to the area's security conditions" (UPI, 2 June 1994). Moreover, the report was rejected almost immediately by Butros Butros-Ghali, the Secretary General of the UN. He said that the United Nations Interim Forces in Lebanon (UNIFIL) had "investigated the claims in the past and had been unable to substantiate them" (Williams 1994, 48). However, the cause, spatial extent, and time of this investigation and its documentation are not known.[3]

ESCWA's follow-up statement is corroborated by Barel (1996), who states that the ongoing war between Lebanon (Hizbullah) and Israel is "the only war conducted by the IDF [Israel Defense Forces] without press coverage. From the Israeli side, journalists are simply not allowed in, and from the Lebanese side Western journalists fear to enter. Lebanese journalists get their reports from the command posts in situ. The IDF and Hizbullah have become their own journalists."

The experience of other researchers is similar to that of the ESCWA field investigators. During the summer of 1992, this writer tried unsuccessfully from both Israel and Lebanon to visit certain Israeli-occupied areas of the Litani River's lower reaches, especially the stretch around its westward bend, to investigate the diversion assertions. Requests for access were denied, apparently when the objective of the visit was declared as being related to the waters of the Litani. In June of the same year, a senior scientist from the National Authority for the Litani River told this researcher in Beirut that he had wanted to scrutinize empirically the widely held belief in an illicit water diversion. He was unable to carry out his research because the Authority's field staff, whose task is to gather data on the flow of the Litani River, were always fired at by soldiers from the Israeli-allied South Lebanon Army (SLA) and the Israeli army, both of which were strategically located on the high grounds on the left (and later on the right) bank of the river. The difficult and dangerous conditions under which research on the Litani River must be done explain the retraction by T. Naff (1993) of his earlier assertion that Israel is trucking water from the Litani. The Israeli armed forces in Lebanon have kept individuals and UNIFIL members away from excavations in critical areas near the Litani's westward bend.

The preceding analyses demonstrate that Soffer's (1993) assertion that researchers can easily check the validity of diversion activities is untrue. This conclusion is based on the comparable experiences of individual researchers on the ground, and of international agencies such as the ESCWA; all were constrained by the protracted state of instability in southern Lebanon. The occupied zone is a militarily insulated area, at least for academic purposes. The question that remains is whether there is convincing substantiation of the theory.

ANALYSES OF EVIDENCE

Supporters of the theory report that Israel is already taking "the water" (al-Husseini 1993) or "some" water of the river (Kassim 1993; Bulloch and Darwish 1993). Others report the diversion to be 20–23 mcm per year (*an-Nahar* 1995; Kassim 1993), 150 mcm per year (ESCWA 1994),[4] or 500 mcm per year (Az-Zu'be 1992; Bulloch and Darwish 1993). These widely varying water figures refer to volumes that are said to be on the verge of or already being diverted. Although most analysts state that the diversion is through a tunnel, some mention a pipeline. Trucking of water from the Litani into Israel was never mentioned by the media or in academic studies in the Middle East as a mode of water diversion.

In 1984 the Lebanese government was concerned about Israel's construction of a tunnel that could "absorb all the water of the Litani River" (*UN Chronicle* 1984, 12), that is to say, 700 to 900 mcm of water per year. These figures are perplexing because of the unrealistically high volume, and because of the immense variation between them and other reported volumes of water that are said to be on the verge of being diverted. Constructing a tunnel large enough to absorb hundreds of mcm of water would require massive movement of earth and hence be all too visible to residents. It must be said, however, that villagers have reported seeing giant construction activities in pivotal areas (believed to be for water diversion). Regarding the volume of water that is alleged to be diverted, the evidence here is sometimes inconsistent and the data unrealistic. This may be due to sloppy research, which is partly explained by the inaccessibility of the region. The researcher N. Khaleefeh states, for example, that water diversion need not be apparent. He argues that it is being done through subterranean channels, and in such a way that effects on the downstream flow of the river are not noticeable (A'yash 1995). However, this assertion is contradicted by M. B. Abboud, who states that the diversion had resulted in "non-seasonal falls in the water level" downstream (Abboud 1986, 4). The essence of Khaleefeh's claim is that small amounts are being diverted. This is consistent with other research findings, which put the diverted volume at 32 mcm (Salmi 1997), and is very plausible from an engineering and hydrological point of view.

The official position of the Lebanese government recently shifted from "more research is needed" to substantiate the diversion theory to "there is no diversion" of Litani waters. Both of these positions are, however, inconsistent with (1) beliefs held by many elected and unelected Lebanese officials and notables from different religious and political persuasions and (2) a recent research report issued by Lebanon's Parliament entitled "Israeli Covetous Interests in Lebanon's Water" (*an-Nahar* 1995). The report quotes American journalists as confirming that Israel has already started to divert 0.250 mcm of water per year through a buried pipeline. The study also asserts that an Israeli-bored tunnel from the Khardali Dam area to Tell al-Nahas is being used to divert water to the Jordan River. The same report quotes the head of Israel's Mekorot water company as having said on 19 July 1990 that his country has been pumping 20 to 23 mcm of water per year from the Litani since 1989 (see also Kassim 1993, 41). Finally, the ESCWA (1994) report is quoted in the same parliamentary study as yet more evidence that Israel is actually diverting the waters of the Litani (*an-Nahar* 1995, 9). Recently, Ash-Shamy (1996, 162) wrote an article in which he concurs with a study by A. Dakrub, who states that "Israel has placed water pumps 1000 meters south of the Khardali bridge and these pumps are connected with huge underground pipes that stretch for 10 kilometers to the water reservoirs of al-Taybih. . . . A seventeen kilometer–long tunnel was also constructed between 1984 and 1986 starting from the Khardali region near Dayr Meemas reaching the border town of Kfir Kala. The pipes are expected to carry one-quarter of a mcm of water to Israel per year." Lebanon's Minister of Information wrote the foreword to his ministry's journal, in which Ash-Shamy's article appeared.

In June 1998, a Lebanese Minister of State, Michele Eddeh, said in a keynote address to an international water conference at the Kaslik University (Jounieh, Lebanon), that Israel is breaking international law by "stealing large amounts of the Litani River's water" from (a neighboring) sovereign republic. When I interviewed this veteran Christian politician, who was close to the President of Lebanon, he spoke of Israel's historical interest in the Litani, and asserted that Israel is currently "stealing water" from the Litani. His evidences are very similar to the ones discussed in this chapter. Neither the Lebanese officials who told me that water is being currently diverted from the Litani nor those who contradicted them had any conclusive or convincing evidence to support their position. This uncertainty and doublespeak are related to the immense scarcity of national water data in general, and more so of data from southern Lebanon, due to Israel's occupation of that region. To be sure, widely conflicting opinions and beliefs about the history and current state of water resources in the Middle East abound (Amery 1997). The Lebanese government needs to resolve the confusion about the Litani River and speak with one voice.

Only two, or perhaps three, Western academics appear to have been cleared by the Israeli army to visit the occupied portions of the Litani River. Although the nature and spatial extent of their field investigations are not published, some have simply stated that the Litani is not being diverted. Such conclusions appear to be influenced by the visually obvious (but surmountable) engineering challenges of diverting water from the Litani gorge to a tributary of the Jordan River. The SLA and the Israeli army's strict control over the Hasbani valley and the security zone, coupled with sporadic fighting in and along the border of the zone, makes it difficult for researchers, journalists, and UN observers to carry out unobstructed field research to conclusively ascertain the nature of Israeli engineering activities. One can conclude that the media and general public in Lebanon, who experience occupation by their longtime enemy, are understandably skeptical of foreign and even space-based technological evidence.

Scientists are rapidly realizing that "some aspects of the traditional scientific method ('We're scientists; trust us') will be misread by a skeptical public as secrecy and obfuscation" (Till 1995, 473). This skepticism partially explains the prolonged life of the water diversion theory, and the actively antagonistic relations that Lebanon and Israel have had for over two decades. Irrespective of evidence and arcane academic debates, it must be realized that the Lebanese public accepts popular, political, and academic reports which support the presence of an Israeli diversion scheme on the Litani River. This poses a clear obstacle to the peace process and to future peaceful relations between the two states. This widely held belief may be dealt with through community- and people-based (collective) confirmation of reality. Public inquiries are normally held because of general concern felt by the community on some issue (Gilpin 1986).

PUBLIC PARTICIPATION FOR PEACE BETWEEN LEBANON AND ISRAEL: A FORWARD LOOK

In addition to scanning their environs for alarm-triggering cues, people tend to scan them for signs of security and safety. A method for assuring the people regarding an environmental concern such as the security of a country's natural resources is through involving the public in the decision-making process. Public input may be obtained through various methods that range from referenda to riots, violence, or threats of violence. The latter result from "frustration with the failure of the other processes available" (Thompson 1984, 188). Public participation is about consulting the people in the deliberations leading to important environmental planning and development decisions. Such a process requires contributions from "the public" (see inset "Defining the Public"). Despite this, some members of the public will reject, resist, or oppose this process. Public participation in resource planning is about

- identification of the effects of a project, and data sources to be used,
- prediction and measurement of the magnitude of impacts and relative objectivity/subjectivity of data,
- interpretation by developing explicit criteria to determine impacts and risk, and
- communication of key issues to the people. This will help them break from their boundedness, and give them a sense of empowerment.

Public participation can become a cementing agent of peace within Lebanon, and between it and its southern neighbor. "Since wars begin in the minds of men, it is in the minds of men that the defenses of peace must be imagined" (UNESCO, as quoted by Eldridge 1986, 208).

The diversion theory can be dealt with indirectly through a public hearing for an irrigation project in southern Lebanon (a branch of which will address the diversion theory), or directly through a fact-finding inquiry regarding the diversion theory. The former will not deal with the theory alone nor at length, nor is it likely to attract sufficient media coverage. It can, however, convince a certain segment of the public. This indirect approach may be preferable because the latter may give the theory a higher-than-needed profile. Either way, the inquiry should seek to, among other things, predict the socioeconomic, political, and environmental impacts of existing or proposed water diversion from the Litani to Israel or elsewhere in the region. If it is not carried out in an open

---

DEFINING THE PUBLIC

THE PUBLIC IS A complex aggregate representing a range of public interests in the region or community affected by the project. The heterogeneous public includes:

- The total adult residents, including those living near a project, and the public at large.
- Regional and local government officials.
- Grassroots organizations, such as homeowners' associations, local environmental groups, farmers' organizations, and the like.
- Professional and business associations.
- Small-business operators.
- Educational institutions.
- Public interest groups.
- The media.

and transparent way, the public participation process may aggravate suspicions and generate greater distrust. Moreover, the process has been historically associated with "extremism, confrontation, delays and blocked development." It can, however, "be used positively to convey information about a development, clear up misunderstandings, allow a better understanding of relevant issues and how they will be dealt with, and identify and deal with areas of controversy while a project is still in its early planning phases" (Glasson et al. 1994, 144–145).

A genuinely independent agency should be set up to coordinate the procedure of the inquiry. The agency ought to make determined efforts to involve the public in the various procedures. It should provide notice of hearings or public meetings, and readily comprehensible water- (or diversion-)related technical documents. If the agency deemed the issue to be of national significance, it should advertise in various media, including official government publications, and write directly to national organizations who may be reasonably expected to be interested in the question. Meetings should be held in various parts of Lebanon, especially in the South or Biqaa provinces. The person conducting the inquiry must ensure that the venue for meetings is convenient, that all parties are given a fair hearing in a congenial atmosphere, and that no testimonies are subjected to rules of evidence, oath, or cross-examinations. Public submissions may be made orally or in writing, and needy persons or groups should be given some level of funding or access to independent technical support to prepare their statements.

Given the media coverage of the diversion theory, the level of consciousness is not apt to be low. Moreover, although there is no tradition of public inquiries in Lebanon, the process is likely to receive acceptance because of the well-established spirit of democracy (despite the infractions and the hitherto limiting presence of foreign forces) that lives on in the country. Public input may provide innovative and viable ideas which, if implemented, would be supported by the public. This would mollify potential escalation in distrust and frustration. The government would gain a much-needed legitimacy—especially in the South—if it uses the process to show that the public is being heard and will be listened to.

SUMMARY AND CONCLUSIONS

This chapter shows that six major historical and contemporary developments have helped in affixing the theory of water diversion in the minds of people in Lebanon, the Arab world, and beyond. One, since at least 1916, Zionist and Israeli officials or senior military officers have expressed interest in the waters of the Litani River. Two, Israel's army has occupied since 1978 the long-coveted water-rich territory. Three, it carried out scientific surveys of potential diversion sites. Four, Israeli scientists repeatedly advance proposals for "cooperative" water diversion from Lebanon. Five, Israeli scientists concur that it is feasible to divert water from the Litani River by "gravitation through a tunnel" (Ben-Shahar et al. 1989, 105). Finally, corroborating evidence of the theory from politicians, academics, and journalists continues to surface. Collectively, these factors give credibility to, though they do not conclusively substantiate, the diversion theory. Therefore, this chapter argued that:

- It is *feasible* to divert water from the Litani into the Jordan River. This deflates the sometimes mentioned notion that the topography near the westward bend of the Litani River creates an insurmountable obstacle to diversion.
- Israel has clear *desires* to utilize the waters of the Litani, and perhaps of other rivers in Lebanon.
- Based on available evidence, it is unlikely that substantial amounts of water are being diverted from the Litani River into the state of Israel. It is, however, very difficult to deny conclusively the existence of minor diversions of water from Lebanon. The diversion of any amount of water from Lebanon into Israel has legal and political implications, not least of which is the internationalization of a national river.

The revocation, dilution, or contradiction of some findings illustrates that some of the diversion allegations have been careless, politically motivated, and/or based on imprecise information. The latter is due to the difficulty of carrying out field research under the existing political-military climate in southern Lebanon. Despite this, and under the current geopolitical and military climates, Israel may be diverting small amounts of water.

As for the future of Lebanese-Israeli water relations, some have convincingly argued that it is "highly probable that Israel will continue to make a determined effort to share or control the waters of southern Lebanon, principally the Litani (or possibly the Awali), in one way or another" (Naff 1993, 17). This will fuel the concern of resource nationalists and of Beirutis who would like to see a greater portion of either river being diverted for their own use. Moreover, and irrespective of its validity, the diversion theory is widely believed in Lebanon, and is frequently reported in bitter tones (e.g., water "theft," Israeli "covetous greed," and the like). Put differently, if solid scientific evidence regarding an environmentally controversial issue were generated from behind military or iron curtains, such evidence would likely be distrusted by those skeptical of government, industry, and administrative or other authorities. Consequently, people's long-held belief in the diversion theory should be dealt with in an inquiry where the public is involved in the decision-making (or fact-finding) process. If the diversion reports are shown to have been bona fide, if such diversions are irreversible (for political or other reasons), or if a negotiated water export arrangement (to Israel or elsewhere) is worked out, the public, specifically in southern Lebanon, must begin to experience, within a short time, the tangible returns of such exports. However, if the theory is shown to be invalid, the open, inclusive process will be one way of sowing seeds of trust (1) between the central Lebanese government and the rest of the population and, significantly, (2) between the Lebanese and the Israelis. The former will help in integrating the multiethnic state, especially if development projects are undertaken to improve the living conditions of the long-neglected southern Lebanese. The latter would be a step toward establishing confidence, and building "warm peace," between peoples on both sides of the last active Arab-Israeli front (many southern Lebanese have relatives in northern Israel).

Although the weight of evidence at this particular juncture is tilted in support of the theory, the contradictions and dilutions of some of the data and findings make it premature to be categorical. Israel's objectives and pursuits in southern Lebanon will continue to be perceived by the Lebanese as menacing due to the bloody history between the two peoples, and due to the geographical positions of prospect and refuge that "the enemy" is in. One may also add that differential endowments in natural resources and power, antagonistic relations, resource interdependence, geographical proximity, and cultural and ideological discrepancies affect the generation of theories of environmental distrust. A definitive statement on the water diversion theory will be possible in due time when pertinent documents become declassified and/or fieldwork in southern Lebanon can go on unhindered. Lonergan and Brooks (1994, 140) state that the "lack of evidence" for the diversion theory does not mean that "some Israelis do not covet the Litani River." Naff (1993, 15) also states that

> There is no disputing Israel's interest in the Litani, just as there is no disputing that Israel is perceived as the most persistent threat to Leb-

anon's rights over the entire flow of the river. All of Israel's neighbors believe that if Israel's water shortage becomes critical enough, Israel would resort to unilateral, arbitrary actions to divert the Litani without regard to international law or censure by the world community. There is no conclusive evidence to that effect, but Israeli behavior has not discouraged such perceptions.

Given Naff's futuristic prediction with respect to Israel's behavior regarding the Litani, and the findings of this chapter, the potential role of public participation in resolving this long-standing theory is promising. Its brief introduction in this chapter is meant to invite further investigation of the potential applicability of this concept. The successful resolution of this diversion theory, especially if carried through with public involvement, will contribute to confidence-building on the domestic and international fronts, and help in developing more responsible water management structures and strategies in Lebanon.

NOTES

1. The theoretical context is very loosely based on Appleton (1975), Archea (1985), Nasar and Fisher (1992), Gates and Rohe (1987), Goffman (1971), and Warr (1990). Consequently, reference is made to specific ideas only.

2. Adapted and molded from Archea (1985).

3. I have failed to even verify the existence of this study, or the one by Buwayz I mentioned. This does not mean that these studies do not exist.

4. According to al-Tadamun (1984), when the pumping stations are put in their place, they will siphon a flow of 150 mcm of water per year into the already constructed tunnel toward Israel.

†Please see postscript p. 115.

REFERENCES

Abboud, Mounir B. 1986. "Lebanon Faces 'River Piracy' Threat from Israel." *Arab News*, 25 August.

AbiNader, Jean. 1998. "The Gulf between the U.S. and Arabs." *Washington Post National Weekly Edition*, 9 March, p. 21.

Agnew, C., and E. Anderson. 1992. *Water Resources in the Arid Realm*. London: Routledge.

al-Ahram. 1992. "Peres Meets a Number of Educated Egyptians." 7 November.

Akins, J. E. 1982. "The Flawed Rationale for Israel's Invasion of Lebanon." *American-Arab Affairs* 2 (Fall): 32–39.

Amery, H. A. 1993. "The Litani River of Lebanon." *Geographical Review* 83 (3): 229–237.

————. 1997. "Water Security as a Factor in Arab-Israeli Wars and Emerging Peace." *Studies in Conflict and Terrorism* 20 (1): 95–104.

————. 1998. "The Role of Water Resources in the Evolution of the Lebanon-Palestine Border." *GeoJournal* 44 (1): 19–33.

————. Forthcoming 1999. Review of *Hydropolitics along the Jordan River: Scarce Water and Its Impact on the Arab-Israeli Conflict,* by Aaron T. Wolf. In *Political Geography.*

————, and A. A. Kubursi. 1994. "The Litani River: The Case against Interbasin Transfer." In D. Collings (ed.), *Peace for Lebanon: From War to Reconstruction,* pp. 179–194. Boulder, Colo.: Lynne Rienner Publishers.

Appleton, J. 1975. *The Experience of Place.* London: John Wiley and Sons.

Arar, A. 1993. "Water Issues in the Palestinian Occupied Territories." Paper presented at the Middle East Water Forum (conference organized by the International Water Resources Association), Cairo, 7–9 February.

Archea, J. C. 1985. "The Use of Architectural Props in the Conduct of Criminal Acts." *Journal of Architectural and Planning Research* 2 (1): 245–259.

Ash-Shamy, A. 1996. "South Lebanon Confronting Israeli Aggression." *Dirasat Lubnaniyah* 1 (Spring): 155–171.

Assi, A. 1968. *Israeli Menace to the Lebanese South.* Beirut: Dar al-Kitab al-Lubnani.

A'yash, A. 1995. "Water in Lebanon's Oil, and Pipelines from the Zahrani to Dahran." *an-Nahar,* 10 August.

Az-Zu'be, A. 1992. *Jewish Invasion of Arab Waters.* Beirut: Dar An-Nafaes.

Baalbaki, A. 1985. *Lebanese Agriculture and the Involvement of the State in Rural Areas from Independence to the Civil War.* Beirut and Paris: Manshuraat Bhir al-Mutawasit.

Barel, T. 1996. "Lebanon: Where We Don't Count Our Dead." *Ha'aretz,* 27 December. English translation in *Middle East International,* 10 January 1997.

Beaumont, P. 1994. "The Myth of Water Wars and the Future of Irrigated Agriculture in the Middle East." *Water Resources Development* 10 (1): 9–21.

Ben-Shahar, H., G. Fishelson, S. Hirsch, and M. Merhav. 1989. *Economic Cooperation and Middle East Peace.* London: Wiedenfeld and Nicolson.

Ben Shaul, D. 1990. "Divvying Up the Drops: Israel Contends with Water Scarcity." *Israel Scene,* June, pp. 5–6.

Bulloch, J., and A. Darwish. 1993. *Water Wars: Coming Conflicts in the Middle East.* London: Victor Gollancz.

Collelo, T. 1989. *Lebanon: A Country Study,* 3d ed. Washington, D.C.: Library of Congress.

Collins, L. 1993. "Sarid Concedes Authorities Hushed Up Radioactive Leak from Negev Reactor." *Jerusalem Post International Edition,* 24 April.

Cooley, J. K. 1978. "Lebanon Fears Loss of Water to Israel." *Christian Science Monitor,* 23 March.

————. 1984. "The War over Water." *Foreign Policy* 54 (Spring 1984): 3–26.

Davis, D. 1990. "Water Shortages Could Lead to War: Scientists." *Jerusalem Post International Edition,* 9 June.

Economist. 1995. "Whose Water?" 5 August, p. 41.

Eisenberg, L. Z. 1994. *My Enemy's Enemy: Lebanon in the Early Zionist Imagination, 1900–1948*. Detroit: Wayne State University Press.

Eldridge, J. 1986. "Public Opinion and the Media." In H. Davis (ed.), *Ethics and Defence: Power and Responsibility in the Nuclear Age*, pp. 45–58. Oxford: Basil Blackwell.

ESCWA (United Nations Economic and Social Commission for Western Asia). 1991. *Environmental Issues in Water Resources and Coastal Management of ESCWA Region*. Arab Ministerial Conference on Environment and Development, Cairo, 10–12 September. Amman, Jordan: Economic and Social Commission for Western Asia (ESCWA), Environment and Human Settlements Division.

————. 1994. *Report on Cooperation among ESCWA Countries in the Field of Shared Water Resources*. E/ESCWA/17/4 (Part I)/Add.3. Amman, Jordan: ESCWA.

E'taani, M. Z. 1984. *The Lebanese-Israeli Agreement and Its Threat to the National Economy*. Beirut: Dar al-Maseera.

FEARO (Federal Environmental Assessment Review Office). 1994. *The Responsible Authority's Guide to the Canadian Environmental Assessment Act*. Hull, Quebec: FEARO.

al-Fursan. 1984. "Israel Is Stealing the Waters of Southern Lebanon with the Aim of Changing the Geography." No. 211, 20 August.

Gates, L., and W. Rohe. 1987. "Fear and Reactions to Crime." *Urban Affairs Quarterly* 22 (3): 425–453.

Gilpin, A. 1986. *Environmental Planning: A Condensed Encyclopedia*. Park Ridge, N.J.: Noyes Pub.

Glasson, J., R. Therivel, and A. Chadwick. 1994. *Introduction to Environmental Impact Assessment*. London: University College of London Press.

Globe and Mail. 1993. "Does NAFTA Let the U.S. Drink Canada Dry?" 19 November.

Goffman, E. 1971. *Relations in Public*. New York: Harper and Row.

Haikal, M. H. 1969. "The Facts in the Middle East Crisis as 1969 Begins." *Al-Anwar*, No. 2939, 3 January.

al-Hawadith. 1984. "Future Drought in Southern Lebanon." No. 1436, 11 May.

Hessler, R. M. 1992. *Social Research Methods*. New York: West Publishing.

Hof, F. C. 1985. *Galilee Divided: The Israel-Lebanon Frontier, 1916–1984*. Boulder, Colo.: Westview Press.

Horowitz, D. 1983. "Israel's War in Lebanon: New Patterns of Strategic Thinking and Civilian-Military Relations." *Journal of Strategic Studies* 63 (3): 83–102.

al-Husseini, J. 1993. "Our Water Security Is in Danger." *Al-Mouhandess* no. 2 (December): 30–36.

Issar, A. A. 1993. "Peace and Development of Water Resources in the Middle East." A poster presentation to the International Symposium on Water Resources in the Middle East: Policy and Institutional Aspects, sponsored by the International Water Resources Association and held at the University of Illinois at Urbana-Champaign, October.

Issar, A. S. 1995. "Climatic Change and the History of the Middle East." *American Scientist* 83 (3): 350–355.

Kally, E. 1991. *Water and Peace: An Israeli Perspective,* trans. to Arabic from Hebrew by Randa Haider. Beirut: Institute of Palestine Studies.

———. 1991–1992. *Options for Solving the Palestinian Water Problem in the Context of Regional Peace.* Israeli-Palestinian Peace Research Project, Working Paper Series No. 19, Winter. Jerusalem: Harry S. Truman Research Institute for the Advancement of Peace.

Kassim, A. 1993. "Interest in Arab Water and Its Geo-political Dimensions." *al-Mustaqbal al-Arabi,* no. 174, 3 May.

Khalifeh, I. n.d., circa mid-1990s. "Israel's Water Strategy with Respect to Lebanon, and Some Practical Proposals for the Lebanese Negotiator." In H. Ramal (ed.), *Water Wealth in Lebanon: Covetous Interests and (Peace) Negotiations,* pp. 36–53. Beirut: Permanent Conference for Lebanese Dialogue.

Kimmerling, B. 1983. *Zionism and Territory: The Socio-territorial Dimensions of Zionist Policy.* Research Paper No. 51. Berkeley: Institute of International Studies, University of California, Berkeley.

Kolars, J. 1993. "The Litani River in the Context of Middle Eastern Water Resources." In J. Kolars, T. Naff, and K. Malouf (eds.), *The Waters of the Litani in Regional Context.* Oxford: Centre for Lebanese Studies.

Le Quotidien de Paris. 1994. "Lebanon Today." 3 June.

Lonergan, S. 1993. "Water and Security in the Middle East." In H. D. Foster (ed.), *Advances in Resource Management,* pp. 199–225. London and Florida: Belhaven Press.

Lonergan, S. C., and D. B. Brooks. 1994. *Watershed: The Role of Fresh Water in the Israeli-Palestinian Conflict.* Ottawa: International Development Research Center.

Lowi, M. R. 1993. *Water and Power: The Politics of a Scarce Resource in the Jordan River Basin.* Cambridge: Cambridge University Press.

Nabeel, Z. 1992. "The Role of the Arab League in the New World Order." *al-Ahram,* 28 November.

Naber, A. A. 1968. "The Arab-Israeli Water Conflict." Doctoral dissertation at the American University, Faculty of the School of International Service, Washington, D.C.

Naff, Thomas. 1990. "The Middle East in the 1990s: Middle East Water Issues." Statement of Thomas Naff in *Hearing before the Subcommittee on Europe and the Middle East of the Committee on Foreign Affairs, House of Representatives,* pp. 152–189. 101st Cong., 2d sess. 26 June. Washington, D.C.: U.S. Government Printing Office.

———. 1993. "Israel and the Waters of South Lebanon." In J. Kolars, T. Naff, and K. Malouf (eds.), *The Waters of the Litani in Regional Context.* Oxford: Centre for Lebanese Studies.

———, and R. C. Matson. 1984. *Water in the Middle East: Conflict or Cooperation?* Boulder, Colo.: Westview Press.

an-Nahar. 1994. "Suspicions in South Lebanon." 24 January.

———. 1995. "A Documentary File from the Parliament in Commemoration of Israel's Invasion of 1978: Israel's Covetous Interests in the Waters of Lebanon under the Microscope." 21 February.

Nasar, J. L., and B. Fisher. 1992. "Design for Vulnerability: Cues and Reactions to Fear of Crime." *Sociology and Social Research* 76 (2): 48–56.

Nasrallah, F. 1990. "Middle Eastern Waters: The Hydraulic Imperative." *Middle East International,* 27 April, pp. 16–17.

O'Riordan, J. 1976. "The Public Involvement Program in the Okanagan Basin Study." *Natural Resources Journal* 16 (1): 177–196.

Rabinovich, I. 1985. *The War for Lebanon, 1970–1985.* Ithaca, N.Y.: Cornell University Press.

Rodan, S. 1996. "Mr. Intelligence." *Jerusalem Post International Edition,* 10 February.

Salmi, R. H. 1997. "Water, the Red Line: The Interdependence of Palestinian and Israeli Water Resources." *Studies in Conflict and Terrorism* 20 (1): 15–65.

Schofield, C. H. 1993. "Elusive Security: The Military and Political Geography of South Lebanon." *GeoJournal* 31 (2): 149–161.

Schwarz, J. 1990. *Israel Water Sector Study: Past Achievements, Current Problems and Future Options.* Draft report, October. Washington, D.C.: World Bank.

Sharif, H. 1978. "South Lebanon: Its History and Geopolitics." In E. Hagopian and E. Farsoun (eds), *South Lebanon,* pp. 9–34. Special Report No. 2. Detroit: Association of American University Graduates.

Soffer, A. 1993. "The Litani River: Fact or Fiction." *Middle Eastern Studies* 30 (4): 963–974.

Stauffer, T. R. 1985. "Arab Water in Israeli Calculations: The Benefits of War and the Costs of Peace." In A. M. Farid and H. Sirriyeh (eds.), *Israel and the Arab Water,* pp. 75–83. London: Ithaca Press.

Stork, J. 1983. "Water and Israel's Occupation Strategy." *Merip Reports* (new name is *Middle East Reports*) no. 116 (July–August): 19–24.

al-Tadamun. 1984. "The Old Israeli Plan Has Begun to Be Realized in Lebanon." No. 54, 21 April.

Thompson, D. 1984. "Avenues of Participation in Natural Resource Development and Disposition." In N. Bankes and J. O. Saunders (ed.), *Public Disposition of Natural Resources,* pp. 18–29. Calgary: Canadian Institute of Resources Law.

Till, J. E. 1995. "Building Credibility in Public Studies." *American Scientist* 83 (September–October): 468–473.

UN Chronicle. 1984. "United States Vetoes Security Council Proposal Concerning Israeli Measures in Lebanon." 21 (21): 11–15.

UPI. 1994. "Explaining a Study Regarding the Litani River." 2 June.

Warr, M. 1990. "Dangerous Situations: Social Context and Fear of Victimisation." *Social Forces* 68 (1): 891–907.

Weisgal, M. W., ed. 1977. *The Letters and Papers of Chaim Weizmann,* vol. 9. Jerusalem: Israel University Press.

Williams, I. 1994. "As U.N. Deals with the Middle East, Anomalies Multiply." *Washington Report on Middle East Affairs,* July–August, pp. 48, 85.

Wolf, A. T. 1995. *Hydropolitics along the Jordan River: Scarce Water and Its Impact on the Arab-Israeli Conflict.* Tokyo: United Nations University Press.

6. *The Water Dimension of Golan Heights Negotiations*

Frederic C. Hof

If and when Syrian-Israeli peace talks centering on the disposition of the Golan Heights resume, the subject of water will, in due course, move to center stage. Although the central focus of both parties will be on military security arrangements, the issue of water presents challenges of its own. Indeed, it is likely that water-related discussions will address not only arrangements for water resources found within Syrian territory conquered by Israel on 9 and 10 June 1967, but issues related to two waterways located to the immediate west of the Golan Heights: the Jordan River and Lake Tiberias.

The purpose of this essay is to clarify the water-related issues faced by the parties and to offer a framework for resolution. On both counts the analysis presented will be largely speculative, because neither side has set forth authoritatively its position, much less its minimal requirements.

Water was an important determinant of the northeastern boundary provided Palestine in 1923 and a significant catalyst for Syrian-Israeli violence between 1949 and June 1967. Should the parties earnestly negotiate the return of the occupied Golan Heights to Syria, the controversy and violence often associated with water over the past seventy-five years will provide the backdrop to some hard bargaining. Yet the issue of water should not, in and of itself, present an insurmountable obstacle to an eventual agreement.

KEY ISSUES

In June 1967 Israel occupied some 1,250 square kilometers of Syrian territory on the Golan Heights, a strategic plateau overlooking the Jordan Valley and lying some 50 kilometers to the west of Syria's capital, Damascus. In October 1973 Syria attempted unsuccessfully to retake the plateau, and the U.S.-brokered Israel-Syria Disengagement Agreement of

31 May 1974 reduced the area occupied by Israel by some 100 square kilometers.

Preliminary Israeli-Syrian peace discussions following the 1991 Madrid Conference focused mainly on potential security arrangements that could facilitate Israel's withdrawal from Syrian territory. A wide sampling of open press reporting on these secret talks leads to the conclusion that no substantive agreements were reached other than an understanding that the Golan Heights, along with strips (the size of which remain to be defined) of Israeli territory to the west and Syrian territory to the east, would in some manner be demilitarized or constituted as limited-forces zones.

Indeed, the question of the *extent* of Israeli withdrawal seems not to have been settled. Prior to the election of the Netanyahu government in Israel in May 1996, the issue appeared to be defined in terms of Syria demanding Israel's "full withdrawal" to positions within or behind demilitarized zones carved out of Palestinian territory by the 1949 Israel-Syria General Armistice Agreement, and Israel responding that the extent of its withdrawal would depend on the adequacy of security arrangements and the nature of the overall peace accord. These preliminary talks ended inconclusively prior to Israel's May 1996 election and have not, as of this writing, resumed.

If they do resume, agreement on security arrangements will continue to be the sine qua non of an overall peace agreement. If the security issue can be successfully tackled, other matters, including cooperative arrangements involving water resources, can be addressed. At this point it appears that the principal water-related issues to be discussed by Israeli and Syrian negotiators are the following:

The Banias Spring/River

One of the three principal sources of the Jordan River (the other two being the Dan River in Israel-proper and the Hasbani River originating in Lebanon), the Banias Spring lies at the base of the northwestern Golan escarpment. Originally assigned to the Palestine Mandate by the 1920 Anglo-French Convention, the spring wound up about one kilometer inside Syria when the border was brought into legal force in 1923. The Banias Spring and the short stretch of the Banias River within Syria came under Israeli control in June 1967 and fall administratively within the occupied Golan Heights. The average annual volume of water provided by the Banias to the Jordan River is 120 million cubic meters (mcm), or about 20 percent of the Jordan's flow as it enters Lake Tiberias. Israel will likely seek assurances from Syria that the Banias will be allowed to flow undiverted into Israeli territory. Syria, which was allocated 20 mcm of the Banias's annual flow for local irrigation by the 1955 Jordan Valley ("Johnston") Plan, may resist making such a commitment.

Golan Surface Water

In order to minimize the impact of Golan Heights settlements on its national water system, Israel has encouraged the construction on the Golan of small reservoirs to impound and distribute (for summer irrigation) winter flood flows that would otherwise find their way either to Lake Tiberias or (via Wadi Ruqqad) to the Yarmuk River.

According to Shalev (1994), there are about a hundred springs of varying yields on the Golan Heights, much of the water from which services Israeli settlements. Shalev also speculates that Syria, should the Golan Heights be returned, could use between 40 and 50 mcm annually of winter flood waters "by using Israeli-built reservoirs" (1994, 162) on the Golan.

Israel will likely seek Syrian assurances that this reservoir network will not be expanded and that steps will be taken to mitigate polluted runoff. Syria would probably resist making any such commitments, but, depending on the tradeoffs involved in an overall negotiated settlement, might countenance undertakings similar to those made by Jordan and Israel in Annex II of their Treaty of Peace pertaining to the environmental protection of a common watershed.[1]

The Jordan River

Although the Jordan does not fall administratively or geographically within the categories of "Golan Heights" or "occupied Syrian territory," there is nevertheless a contentious issue summarized by the phrase "the line of 4 June 1967."

From 1949 until June 1967 the upper course of the Jordan River served as the de facto boundary between Syria and Israel, even though the river fell entirely within British Mandate Palestine. Syria, which was assigned an allocation of 22 mcm annually from the upper Jordan by the 1955 Jordan Valley Plan to irrigate a farm area near Lake Tiberias, may seek to make the river itself the new international boundary. Israel can be expected to resist vigorously the redrawing of the 1923 mandate boundary *unless* the line is moved to the east at Syria's expense.

Lake Tiberias

Lake Tiberias, like the Jordan River, is neither part of the Golan Heights nor a waterway formerly governed by Syria. The 1923 mandate boundary kept the lake entirely within Palestine, placing the line along the northeastern shore a mere 10 meters off the high-water mark.[2] By virtue of an Anglo-French "Good Neighborly Relations" accord enacted in 1926, Syrian fishermen were allowed access to the lake.[3] It is possible that Syria will seek the internationalization of Lake Tiberias, a proposal sure to be resisted by Israel. Access for Syrians may, however, be another matter entirely, especially if the 1923 "10 meter" boundary prevails.

HISTORICAL BACKGROUND

The following narrative seeks only to summarize key points under the following headings: the creation of the 1923 "international boundary"; the 1949 armistice regime and associated problems; the Johnston mediation; Arab diversion efforts; and the water-related implications of Israel's occupation of the Golan Heights.

The 1923 Boundary

The end result of a complex, post–World War I diplomatic process involving Great Britain, France, and the Zionist leadership was a Palestine-Syria boundary which kept both the upper Jordan River[4] and Lake Tiberias a few meters inside Palestine. Although the inclusion within Palestine of the entire Lake Tiberias was absolutely essential for Zionist water planning[5] (Tiberias being the region's natural reservoir), two of the three sources of the Jordan (the Banias and Hasbani Rivers) wound up in Syria and Lebanon, respectively.

The 1949 Armistice

At the end of Arab-Israeli fighting in 1948, Syrian army units were ensconced on several square kilometers of Palestinian territory. Israel, in order to effect their removal, agreed to a UN proposal that those areas occupied by Syrian forces, along with other militarily exposed tracts of Palestinian land east of Lake Tiberias, be designated a "demilitarized zone" (DMZ), whose sovereign identity would be deferred pending a formal peace between Israel and Syria. The DMZ consisted of three noncontiguous enclaves connected to one another by an "armistice demarcation line" which followed the 1923 international boundary.[6] It is worth noting that Israel, notwithstanding the terms of the armistice, considered the entire DMZ to be part of Israel. Syria never asserted a parallel claim.

According to Muslih (1993, 614), "Israel's creeping annexation of the DMZ and Syria's determination to check Israeli advances dominated much of the history of the 1949–1967 period." Israel claimed that there was nothing to annex, that the land in question had been Palestine (assigned to the Jewish state by the 1947 UN partition plan) and was now Israel. Syria did not claim sovereignty over the zone, but held that Israel's unilateral claim was invalid, and that Israeli actions within the DMZ— especially the draining of the Huleh Lake marsh and the attempt to divert Jordan River water from within the DMZ—were at variance with the General Armistice Agreement.

Israel eventually relented on the DMZ diversion issue (moving the diversion point to Lake Tiberias), but repeated acts of violence on and near Lake Tiberias and all points along the DMZ from Dan in the north to El Hammé in the south characterized this troubled period. According to Muslih, "Israeli tractors would move into disputed areas, often with the support of armed Israeli police. The Syrians would fire from their high ground positions, and would often shell Israeli settlements in the Huleh Valley. By trying to oppose the Israeli challenge, Syria drew on its head punitive Israeli raids, including air strikes" (Muslih 1993, 619).

Today's legacy of the armistice regime is the "line of 4 June 1967" controversy. At issue is whether to define Israel's "full withdrawal" in terms of the 1923 international boundary (which would formally convey Syrian recognition of Israeli sovereignty over the 1949 DMZ), or a line that somehow takes into account the armistice regime and the tactical situation "on the ground" just prior to Israel's conquest of the Golan Heights. In demanding a full Israeli withdrawal to a line west of the 1923 boundary, Syria would emphasize that the upper Jordan River was the de facto boundary between Israel and Syria until the June 1967 war[7] and that Syrian forces were actually on the shores of Lake Tiberias.[8] Despite the overall success of Israel's "creeping annexation" campaign in the

65-square-kilometer DMZ, Syria itself controlled some 18 square kilometers of the zone on the eve of war in June 1967 (*Ha'aretz* 1995).

The Johnston Mediation

With respect to the Golan Heights and adjacent areas, the Jordan Valley ("Johnston") Plan incorporated the recommendations of the Arab League Technical Committee and allocated to Syria 20 mcm from the Banias River for local irrigation and 22 mcm from the upper Jordan River to service "the irrigable area in the vicinity of Boteiha Farms" (U.S. Department of State 1955, 13) in Syrian territory near Lake Tiberias. The plan also envisioned "A new diversion structure and canal from the Jordan River to Boteiha Farm in Syria, together with 50 K.W. of electric power to replace water power" (U.S. Department of State 1955, 12).

The proximity of the Banias Spring and the Syrian portion of the river to disputed territory in the DMZ helped to make the 20 mcm allocation to Syria essentially meaningless.[9] Neither did the Boteiha Farm area receive the benefits of the water (22 mcm annually) and infrastructure cited in the plan. It is beyond dispute, however, that an American-led mediation effort whose validity and relevance has, at times, been upheld by Israel,[10] allocated 42 mcm of water to Syria from sources very likely to be the subjects of negotiation should Israel and Syria return to the bargaining table. Furthermore, the Jordan Valley Plan allocated from the upper Jordan/Lake Tiberias 100 mcm (85 mcm of this amount of high quality, the balance saline) to the Hashemite Kingdom of Jordan (East and West Banks), a transfer never effected.

Will Syria raise the issue of Jordan Valley Plan allocations in negotiations with Israel? It is worth noting that Syria has more than compensated itself for the "loss" of Jordan and Banias water at *Jordan's* expense, by depleting the Yarmuk River at a rate of 220 mcm per year,[11] as compared to the plan's allocation of 90 mcm—the latter figure recommended by the Arab League Technical Committee and accepted without dispute by Ambassador Eric Johnston (U.S. Department of State 1955, 13). El-musa (1996, 33) has noted that "Syria has not taken a public stance regarding the plan during the peace negotiations. Assuming it regains the Golan Heights, it would probably rest its demands, as upstream riparians usually do, on its sizeable contribution to the system." This stand would be consistent with Syria's upper riparian position vis-à-vis Jordan, but not its lower riparian position with respect to Turkey and the Euphrates River.

The Arab Diversion Plan of 1964

Israel's diversion of water from Lake Tiberias was the catalyst for a January 1964 Arab League summit in Cairo. Annoyed by constant Syrian charges that Egypt was less than eager to confront Israel militarily, Egypt's President Gamal Abdel Nasser championed a diversion scheme aimed at routing two sources of the Jordan River (the Hasbani and the Banias) away from Israel (Kerr 1971). According to a contemporary account, "The total amount of water that the Arabs plan to use amounts to 77 per cent of the Jordan-Yarmuk system. . . . This would leave Israel with the remaining 23 per cent, rendering the Israeli water project useless and preventing it from irrigating the Negev" (Saliba 1968, 110).

Israeli air and artillery strikes on Syrian heavy equipment stopped the preliminary phase of the diversion scheme dead in its tracks by July 1966 (Shalev 1994), nearly a year before the occupation of the Golan Heights. The diversion crisis had added fuel to the fire, but the threat and fact of Syrian shelling from the high ground, not a desire to annex a water source, was the decisive factor in Israel's decision to seize the Golan Heights. In the words of Shimon Peres, "The Golan Heights was never historically considered a part of the State of Israel. Had we not been attacked from the Golan Heights we would never have gone up there" (Jerusalem, Israel Television Channel 1, Network 2, 2 June 1995).[12]

Water-Related Implications of the Occupation

On 9 and 10 June 1967 Israel captured some 1,250 square kilometers of Syrian territory, including the Banias source, the Golan Heights, and the Golan's principal town, Quneitra. Renewed fighting in October 1973 led to the Israel-Syria Disengagement Agreement on the Golan Heights of 31 May 1974, which returned Quneitra to Syrian control and reduced the size of Israel's holdings to about 1,150 square kilometers.

During the 1967 fighting Israel also expelled Syria from the entire 1949 DMZ, including the El Hammé salient along the Yarmuk River. When El Hammé, a Palestinian Arab town across the Yarmuk from Jordan, became the Israeli town of Hamat-Gader, Israel's frontage along the Yarmuk River was extended significantly. An important water-related consequence of the 1967 war was that Jordan had a new neighbor right across the stream from the intake of its East Ghor Canal, the vital conduit for Yarmuk waters irrigating the Hashemite Kingdom's share of the Jordan Valley.

The major water-related implications of the June 1967 war on the Israeli-Syrian front were three: the Banias Spring and River came under complete Israeli control; Syrian forces were expelled from the vicinities of the upper Jordan River and Lake Tiberias; and Israeli settlements, with water needs of their own, began to appear on the Golan Heights.

Yet potentially the most important hydro-political consequence of war and occupation was the growth in Israel of the idea that the water resources seized in June 1967 should be kept under Israel's sovereign control, a proposition that would obviously and perhaps decisively affect negotiations with Damascus, which has made Israel's full withdrawal from Syrian territory the minimum price of peace. Equating "water security" with sovereignty likewise has important implications for Israeli-Palestinian "Final Status" negotiations.

In 1991 Tel Aviv University's Jaffee Center for Strategic Studies commissioned two researchers to undertake a comprehensive study of the regional water situation and the prospects for multilateral cooperation. Publication of the report was subsequently blocked, reportedly because it contained maps "outlining possible lines of withdrawal from the West Bank and the Golan Heights that would still safeguard Israel's water sources" (Schiff 1993). A map accompanying an analysis prepared by the Washington Institute for Near East Policy depicted a Jaffee Center "water security line" on the Golan Heights which could form the basis of a new boundary, a line which "by and large passes not far from the 1967 lines, aside from the southern portion of the heights, east of the Sea of Galilee" (Schiff 1993). The Jaffee Center's "water security line" would presumably secure two objectives: incorporate the Banias Spring into Israel-proper, and annex to Israel the eastern approaches to Lake Tiberias and the heights draining into it. The balance of the Golan Heights could, according to the report, be returned to Syria.

The report was officially suppressed because it indeed suggested that some territory might be returned to Syria without causing a negative hydrological impact on Israel, and that even more might be returned provided adequate safeguards and compensation were built into an agreement. Evidently the report's impact, suppressed or not, survived the passing of Likud and the arrival of Labor. According to an Israeli press report in July 1995, Prime Minister Yitzhak Rabin told a group of Israeli ambassadors that "the greatest danger Israel has to face in the negotiations with Syria is the possibility of losing control over the Golan Heights' water resources" (*Ma'ariv* 1995).

Israel's concerns about the water implications of a Golan Heights withdrawal reflect, perhaps, one of the great ironies of the Arab-Israeli dispute. Thanks in large measure to the constant tension and periodic violence of the 1949–June 1967 era, a militarily superior Israel had its way almost completely vis-à-vis Syria on water issues. It forced Syria to allow the Banias to flow freely into Israel, it kept Syria from exploiting the Jordan River, and, except for the occasional presence of fishermen, it prevented Syria from using Lake Tiberias. Now Israel may be faced with a new situation: a Syrian government secure enough (albeit on the basis of total Israeli withdrawal) to make peace. Controversies involving water which were solved unilaterally by force and occupation may now be subject to negotiation, a process inevitably entailing compromise. Peace may be its own reward, but a country long preoccupied by water may find it hard to bargain over resources sought by the founders and secured by their grandsons.

OUTLINE OF A WATER SETTLEMENT

Although it is still possible to envision a deus ex machina diplomatic breakthrough engineered by American diplomacy, one involving perhaps a major Turkish concession to Syria on the Euphrates in return for Syrian water-related guarantees to Israel, the discussion which follows assumes no such fortuitous event. Neither does it assume that the United States and others will step in with generous financing for desalination and other technological "fixes" which might encourage Israelis to discount their fears of a Syrian hand on the faucet, although international assistance to both parties along those lines might be appropriate and important.

It is assumed, on the contrary, that Israel and Syria will be obliged to address these matters without the benefit of a bailout, that no one, save the parties themselves, will "save the day."

It is also assumed, for the sake of simplicity, that Israel's prospective withdrawal will be defined as "complete" in terms of the 1923 international boundary; that the "line of 4 June 1967" controversy will be solved by folding the entire 1949 DMZ into a new DMZ or an area of limitation in armament and forces; and that all territory which was indisputably Syrian prior to the June war will be returned to Syria. Such an outcome would uphold, in an honorable manner, the intent of the 1949 General Armistice Agreement, which was one of disengagement and demilitarization. Syria, which has never asserted sovereignty over any part of Palestine, may well, in the context of a comprehensive agreement, find such an arrangement acceptable. If, however, Damascus chooses to claim pieces of Mandatory Palestine, it may find itself at odds not only with Israel, but with the Palestine National Authority (the only entity other than Israel currently asserting claims of sovereignty to the former Mandatory Palestine).

The parties will, no doubt, come to the negotiating table with maximalist positions. Israel will likely argue that Syrian use of water on and near the Golan Heights should be tightly circumscribed, both in terms of quantities and environmental protection, and that Syrian access to the Jordan River and Lake Tiberias is a nonstarter. Syria will likely open with the position that its use of water resources inside Syria is not something to be subjected to outside control or even monitoring. Syria may also insist that Israel recognize the right of Syrian citizens to enjoy access to both the upper Jordan River and Lake Tiberias, either by internationalizing both waterways or by instituting access arrangements short of challenging Israeli sovereignty.

The negotiating positions of both parties may likewise be influenced by regional political considerations involving the issue of water. Israel faces protracted "Final Status" negotiations with the Palestine National Authority over water, talks which may well review those provisions of the Jordan Valley Plan calling on Israel to provide 100 mcm (85 mcm of high-quality water) to the pre–June 1967 Hashemite Kingdom of Jordan. Likewise Israel and Jordan may, when Israel-Syria negotiations commence, still be facing difficulties in implementing the water-related provisions (Annex II) of their 1994 peace treaty, one of which involves an Israeli commitment to help Jordan obtain 50 mcm per year in additional drinking water. Finally, Syria's massive and probably illegal depletion of Yarmuk River headwaters has done substantial harm to Jordan, even to the extent of making Jordan's prospective "Unity Dam" (agreed to by Syria in a 1987 treaty) infeasible.[13]

The parties might well analyze their bilateral water issues in this broader regional context with a view toward applying their bilateral undertakings to the resolution or amelioration of other regional water controversies. Such an approach admittedly runs against the grain of normal diplomatic practice, which tends to isolate specific bilateral problems from broader regional concerns and complications. In the specific case of Israeli-Syrian negotiations over the Golan Heights, however, the parties may find that their own efforts to find common ground can produce compromises whose political sustainability might actually derive in part from their usefulness in mitigating other problems as well. Israel and

Syria may, in short, be able to kill many birds with a few well-aimed stones.

By using the 1955 Jordan Valley Plan as a reference point and guide, it may be possible for Israel and Syria to protect their basic equities in a manner which enables the leadership of each state to defend the overall agreement domestically. An examination of the key issues in the context of the "Johnston Plan" illustrates this approach.

The Banias Spring/River

Although the 1955 Jordan Valley Plan alludes to the possibility of irrigating low-lying Syrian lands in the vicinity of the Banias Springs, and generously[14] allocated 20 mcm per year for that purpose, Israel has, since 1967, allowed the Banias to flow freely into the Jordan.

Syria might wish to consider obliging Israel and substantially continue that practice. It could do so by limiting its own Banias withdrawals to purely local uses (thus forgoing altogether the possibility of pumping Banias water up to the Golan Heights) in return for an Israeli commitment to provide some portion of the 20 mcm not used by Syria to *downstream* Arab riparians (Jordan and the Palestine National Authority).

In essence Syria would be assigning part of its "Johnston Plan" allocation to others, trading water for which it may have little objective use except within a very small and isolated corner of Syria in order to make a gesture of solidarity with other Arab parties. Syria's high-handed behavior vis-à-vis Jordan in the Yarmuk watershed, a practice no doubt noted with interest by Turkey, would seem to suggest that an act of statesmanship might be in order. The Syrian Arab government would also be in a position to justify domestically its willingness to forgo, once and for all, the diversion of the Banias by referring to its high-minded defense of broader Arab interests.

Israel's objective with regard to the Banias is to bind Syria to a legal commitment not to undertake diversion projects, to allow the river to flow, for the most part, into the Jordan. Having already agreed in the 1950s to an allocation of 20 mcm from the Banias to Syria, Israel might prove willing to "spend" up to 20 mcm to keep the bulk of the river's average annual flow (100 of 120 mcm) within the river's natural channel.

A secondary objective for Israel might be to convince Syria that an allocation of 20 mcm for Syrian local uses in the Banias village vicinity is excessive; that the "Johnston Plan," while providing a useful point of reference, sought to allocate the waters of the Jordan Valley on the basis of irrigable land, an approach that no longer, in the 1990s, makes economic sense. In the course of bargaining over the size of Israel's obligation to Syria, the parties might jointly conclude that each can gain by designating several mcm of the water in question to flow naturally to Lake Tiberias for distribution to downstream riparians. Syria's potential stake in this approach was examined above. From Israel's perspective, if a quantity of water is to be sacrificed for the sake of an undiverted Banias, some of it may as well be used to help in facilitating downstream water arrangements with Jordan and the Palestine National Authority.

The Jordan River

If the 1923 international boundary prevails as the "peace boundary" between Syria and Israel, the Jordan River will remain entirely within Israel, albeit within 50 meters of Syria at some points. Both sides of the river will likely fall within a new demilitarized zone, and Syria may well insist that its nationals, returning to areas abandoned decades ago, be granted access to the Jordan.

Inasmuch as border security will remain a very sensitive issue between the parties, Syrian access might be implemented in the form of an annual allocation from the river piped to inhabited areas in Syria west of and below the Golan escarpment. The Jordan Valley Plan allocated 22 mcm to Syria from the Jordan River. As discussed above in the context of the Banias, the parties might critically review the assumptions underlying the 1955 allocation and conclude, each for its own reasons, that some part of the 22 mcm be allocated to Syria, with the balance going into Lake Tiberias for distribution by Israel to downstream Arab riparians.

Lake Tiberias

Reinstitution of the 1923 boundary could place Syrian citizens almost literally within spitting distance of Lake Tiberias. Were Syria to drop its long-standing goal of internationalizing the lake, its citizens might be granted reasonable access for fishing. Depending on where returning Syrians actually settle, it may also prove feasible to provide for their water needs from the lake instead of the Jordan River. Access has its precedent in the 1926 Anglo-French Good Neighborly Relations accord, and would seem to be a reasonable alternative to the tantalizing and provocative practice of placing a waterway completely off-limits to people peering at it through a fence 10 meters away from the water.

Furthermore, the Jordan Valley Plan allocation of 85 mcm of high-quality water annually to downstream Arab riparians might, if resurrected, be used to help accomplish the following goals:

1. reconcile Syria to full acceptance of the 1923 boundary, including those provisions that place important water resources just beyond its sovereign reach;

2. facilitate implementation of Annex II provisions of the Israel-Jordan peace treaty, particularly Article 1, paragraph 2, point 3, which obligates the parties to "cooperate in finding sources for the supply to Jordan of an additional quantity of fifty MCM/year of water to drinkable standards";

3. facilitate progress in Israeli-Palestinian "Final Status" talks on water; and

4. establish a basis for the cooperative, multilateral management of water in the Jordan Valley watershed, focusing on the region's natural reservoir: Lake Tiberias.

Golan Surface Water

The parties will be obliged to deal with two issues related to water originating on the Golan Heights: prospective Syrian use of that water for agricultural and municipal purposes, and the potential for polluted runoff to enter Israel's water system.

As noted above, springs on the Golan currently serve the needs of Israeli settlements, and reservoirs built by Israel on the Heights can impound 40–50 mcm annually (Shalev 1994). Shalev also believes that Syria could feasibly impound another 30–40 mcm annually by damming wadis. He does not specify, however, how much of this water currently flows into the Jordan and Tiberias as opposed to the Yarmuk, where Jordan would have the first opportunity to capture it.

When the time comes to address the issue of water resources originating on the Golan Heights, the parties will likely open by staking out maximalist positions. Syria will indicate its intention to repopulate the area and will resist the notion that its use of water should be limited in any way. Israel, ideally, would wish to dismantle the reservoir and irrigation network it has built and restrain Syria from rebuilding it, thereby maximizing the amount of water flowing into the Jordan River and Lake Tiberias.

The outline of a potential compromise is not difficult to discern. By establishing settlements on the Golan and investing in the water needs of Israeli Golan residents, Israel has already accepted, in effect, a marginally reduced rate of flow from the Golan into its national water system. Indeed, to the extent that Israel has been obliged to pump water up from Tiberias to service the needs of settlers, this requirement will disappear altogether if the Golan is returned to Syria.

A Syrian commitment not to expand Israel's current water infrastructure on the Golan Heights and not to build upon Israel's current level of water usage on the Heights might, from Israel's perspective, constitute an acceptable status quo arrangement. Whether or not such a commitment by Damascus would be consistent with Syrian resettlement plans is not known.

With regard to potential pollution, the central role of Lake Tiberias in Israel's national water system accounts for Israeli concerns. What is often ignored is that Jordan also has an interest in this matter, as Golan winter floods not captured by Syrian dams in Wadi Ruqqad make their way into the Yarmuk River for the potential use of Jordan and, farther downstream, Israel.

Syria might well acknowledge in the peace treaty its special responsibility as the upper riparian for the environmental protection of watershed areas under its control, and might undertake to enforce that protection in accordance with mutually agreed standards. The treaty might also provide (as does the Israel-Jordan treaty) for a joint water committee that would monitor environmental matters.

CONCLUSION

As of early 1998, the sine qua non of formal peace between Syria and Israel is, from Syria's perspective, the full withdrawal of Israel from Syrian land occupied during the June 1967 war. Moreover, Syria has taken the position that Israel should withdraw to the "line of 4 June 1967," a formulation which, at the very least, implies the removal of Israeli forces from the 1949 DMZ, or its partition.

The current government of Israel has manifested, as of early 1998, no interest in negotiating with Syria on the basis of "full withdrawal," whether to the 1923 international boundary or to the line of 4 June 1967. A task for American diplomats in the second Clinton administration will be to assist Israeli officials in determining whether or not there might potentially be conditions under which Israel might actually countenance a "full withdrawal." If the government of Israel takes the position that under no conceivable circumstances would it withdraw from all Syrian territory occupied since June 1967, then there would appear to be no basis for Israeli-Syrian peace negotiations unless the Syrian Arab government were to indicate its readiness to negotiate on the basis of something less than full withdrawal, an unlikely prospect.

It is possible that the parties might agree to negotiate on the basis of full withdrawal and still not achieve closure. The principal ingredient in the success or failure of such negotiations will be the attainment of a security regime mutually acceptable to Israel and Syria.

If the parties are able to devise acceptable security measures in the context of a full Israeli withdrawal, differing perspectives on the water issue should not present an insurmountable obstacle to a full and final peace agreement. Indeed, Israel should be able fully to preserve its access to the upper Jordan watershed consistent with the Jordan Valley Plan and to secure Syria's agreement to the environmental protection of Lake Tiberias and the Yarmuk River. Syria, in return for acknowledging Israeli sovereignty over the upper Jordan River and Lake Tiberias, might receive access to those waterways in ways consistent with the 1926 Good Neighborly Relations accord and the Jordan Valley Plan. To the extent that the "Johnston Plan" may have incorporated an inflated estimate of Syrian water requirements in the Jordan Valley, Syria may wish to consider "assigning" some of its allocation to Arab lower riparians, thus enabling those riparians to achieve closure with Israel on water-related provisions of other treaties and agreements.

Ultimately the riparians of the Jordan Valley—Israel, Syria, Jordan, and Lebanon—may come to recognize the validity of Ambassador Eric Johnston's central point: that multiple sovereignties within a watershed should eventually be subordinated to multilateral, but unified, management, perhaps on the basis of a succession of bilateral agreements. Israel and Syria, in trying to overcome a legacy of hostility and distrust to establish formal peace between them, must first agree on security measures that neutralize that legacy and lay the basis for a different sort of relationship. If they succeed in that endeavor, there is nothing in the way of water disputes that should prevent them from signing and implementing a treaty of peace.

NOTES

Author's note: This essay, in substantially the same form, first appeared in *Middle East Policy* 5 (2): 129–141.

   1. Article 3 of Annex II addresses "Water Quality and Protection."

   2. According to the 3 February 1922 final report of the Anglo-French

boundary commission, "the frontier follows a line on the shore parallel to and at 10 meters from the edge of Lake Tiberias, following any alteration of level consequent on the raising of its waters owing to the construction of a dam on the Jordan south of Lake Tiberias" (Toye 1983, 3:442).

3. The inhabitants of Syria and Lebanon were granted the same fishing and navigation rights on Lake Huleh, Lake Tiberias, the Jordan River, and all related watercourses as the inhabitants of Palestine, but "the Government of Palestine shall be responsible for the policing of the lakes" (Toye 1983, 3:443, 497).

4. The final boundary placed the Jordan River within 50 meters of Syria along one stretch just to the north of Lake Tiberias (Toye 1983, 3:442). "From Lake Huleh to the Lake of Galilee [Tiberias] the boundary runs a short distance, varying from 50 to 400 yards, east of the Jordan River and almost parallel to it" (Brawer 1968, 236).

5. "... the intention of the boundary makers in 1922 was to grant Palestine full legal ownership of the Jordan and its lakes so that there should be no necessity to obtain the consent of any other country for any project to utilize the waters of the river" (Brawer 1968, 237). This may indeed have been the intention of the *British* "boundary makers," but their *French* counterparts succeeded in placing two key sources of the Jordan in Lebanon and Syria, respectively. Israel has never accepted the proposition that its neighbors have the right to do as they wish with the Hasbani and Banias by virtue of "full legal ownership."

6. Article V.1 of the General Armistice Agreement specifies that "arrangements for the Armistice Demarcation Line between the Israeli and Syrian armed forces and for the Demilitarized Zone are not to be interpreted as having any relation whatsoever to ultimate territorial arrangements affecting the two Parties to this Agreement." Article V.2 defines the purpose of these arrangements as "separating the armed forces of the two countries ... while providing for the gradual restoration of normal civilian life in the area of the Demilitarized Zone, without prejudice to the ultimate settlement." Article V.5(a) states that "Where the Armistice Demarcation Line does not correspond to the international boundary between Syria and Palestine, the area between the Armistice Demarcation Line and the boundary, pending final territorial settlement between the Parties, shall be established as a Demilitarized Zone. ..."

7. "The narrow strip east of the Jordan was never in fact controlled by the British Mandatory Authorities, nor has it been by the Israelis" (Brawer 1968, 237).

8. "The fact that the Syrians actually hold small strips of Palestinian territory in the upper Jordan Valley has brought sections of the eastern bank of the Jordan and shores of the Lake of Galilee [Tiberias] under their control" (Brawer 1968, 238).

9. "So far [on the eve of the 1967 war], almost the entire waters of the Banias have been reaching Israel, only a very small part being used in Syria" (Brawer 1968, 229).

10. Most recently by Minister of Agriculture Rafael Eitan on 27 August

1996. In commenting on Jordanian-Syrian water discussions, Eitan insisted that the distribution of Yarmuk River water be in accordance with the Johnston Plan (*Jerusalem Post* 1996).

11. The figure of 220 mcm is according to Dr. Munther Haddadin, Jordan's water negotiator with Israel (*Jordan Times* 1994).

12. Shortly after the end of the June 1967 war, the government of Israel took the policy position that Israel could withdraw from the Golan Heights to the 1923 international boundary in exchange for full peace with Syria.

13. On 13 October 1997 the Jordan-Syria Joint Commission, charged with overseeing the 1987 treaty governing the construction of the Unity Dam, agreed that the project should go forward once Jordan updates all relevant studies, "taking into consideration the rising cost of construction these days and the drop in the level of the river water since 1987" (*Jordan Times* 1997). The "drop" referred to above will likely mandate a significantly smaller dam than the one envisioned in 1987, perhaps located well downstream of the original site.

14. "The allocations to Lebanon and Syria are all within the upper basins, and water which is diverted but not actually consumed will return to the river system, either by flow underground or surface run-off. *Thus, while the allocations may be liberal, any excess will not be wasted since it may be re-used downstream*" (U.S. Department of State 1955, 13–14 [emphasis added]).

REFERENCES

Bar-Yaacov, N. 1967. *The Israel-Syrian Armistice*. Jerusalem: Magnes Press.
Brawer, Moshe. 1968. "The Geographical Background of the Jordan River Dispute." In Charles A. Fisher (ed.), *Essays in Political Geography*, pp. 225–242. London: Methuen & Co., Ltd.
Cook, Steven A. 1995. *Jordan-Israel Peace, Year One: Laying the Foundation*. Research Memorandum No. 30. Washington, D.C.: Washington Institute for Near East Policy.
Elmusa, Sharif S. 1996. *Negotiating Water: Israel and the Palestinians*. Washington, D.C.: Institute for Palestine Studies.
Ha'aretz. 1995. "Syria Demand for Withdrawal to Armistice Lines Reviewed." 29 May.
Hof, Frederic C. 1997. "The Water Dimension of Golan Heights Negotiations." *Middle East Policy* 5 (2): 129–141.
Israel Television Channel 1, Network 2. 1995. Interview with Foreign Minister Shimon Peres by political correspondent Gadi Sukenik at the Foreign Ministry in Jerusalem. 2 June.
Jerusalem Post. 1996. "Eytan—U.S. Must Insist Jordan, Syria Adhere to Accord." 28 August.
Jordan Times. 1994. "Syria Said 'Diverting' Water from Country's Share." 3 December.
———. 1997. "Jordan, Syria Agree to Build Dam on al-Yarmuk River." 14 October.

Kerr, Malcolm H. 1971. *The Arab Cold War: Gamal 'Abd al-Nasir and His Rivals, 1958–1970,* 3d ed. New York: Oxford University Press.

Ma'ariv. 1995. "Rabin: Water 'Greatest Danger' in Syria Talks." 19 July.

Muslih, Muhammad. 1983. "The Golan: Israel, Syria and Strategic Calculations." *Middle East Journal* 47 (4): 611–632.

Saliba, Samir N. 1968. *The Jordan River Dispute.* The Hague: Martinus Nijhoff.

Schiff, Ze'ev. 1993. "Israel's Water Security Lines." *Policywatch.* Peace Watch Number Seventy-Five. Washington Institute for Near East Policy, 4 November.

Shalev, Aryeh. 1994. *Israel and Syria: Peace and Security on the Golan.* Tel Aviv: Tel Aviv University, Jaffee Center for Strategic Studies.

Toye, Patricia, ed. 1983. *Palestine Boundaries 1833–1947,* vol. 3. Oxford: University of Durham Press.

U.S. Department of State. 1955. "The Johnston Plan." Typescript memorandum. 30 September.

7. *Water Security for the Jordan River States*
PERFORMANCE CRITERIA
AND UNCERTAINTY

Paul A. Kay and Bruce Mitchell

INTRODUCTION

The question of designing a sustainable future moved onto the global agenda with the United Nations Conference on Environment and Development in June 1992. In its classic formulation, sustainable development was defined as utilization of resources now in a manner that would not preclude options for future generations (World Commission on Environment and Development 1987). Growing concerns about global change, including loss of biodiversity and climatic change, raise questions of whether development can be sustainable. New ways of recognizing that environment is variable and that the future is uncertain are required for policy-making and long-range planning. In regions where a resource essential to life is in short supply, the issues of variability and uncertainty take on immediate importance. Such a region is the Jordan River basin, and such a resource is water.

Much has been written about the importance of water to the peace process in the Jordan River basin region, mostly emphasizing the political framework. Water scarcity may exacerbate political, economic, and military power imbalances, or may offer opportunity to lead or closely follow settlement of political issues (Lowi 1992; Gleick 1994; Lonergan and Brooks 1994; Wolf 1995). This emphasis on political analysis, however, has the effect, paradoxically, of making the environmental conditions—scarcity, variability, uncertainty—a fixed background against which the state drama is played. Many proposals exist for mega-projects designed to increase the amount of usable water in the region, without reference to natural variability in the region (for example, Fishelson 1995). The emphasis on water rights or technology does not guarantee that sustainability will be achieved. The question of sustainability must include environmental considerations as much as political and economic ones. Brooks (1994), for example, has argued for a soft-path, demand-side approach (in contrast to the technological focus on more water), which would promote conservation and allocations according to economic and environmental quality criteria. We argue for a fuller recognition of supply variability. How well a society has lived with variability may be informative about how well it might live with future change.

Our goal in this chapter is to examine what adjustments to variability of water resources have been made, and then to suggest how management of water may be approached to incorporate issues of variability and uncertainty. We apply some concepts of performance evaluation of systems and of adaptive management to the question of water security in the Jordan River states. Fundamentally, we ask how water resources have been managed to cope with scarcity and variability, and what this implies for a future of coping with uncertainty, both environmentally and socially.

The quantitative analyses that follow, for performance evaluation, focus on Israeli water use, for two reasons: Israel is the largest, most developed water user in the Jordan River basin, and arrangements for shared resource management will be made in this context; and, time series data are readily available and of good quality for Israel, but not for other states or entities in the region. Israel's experience, and the place of water in national policies and in the peace instruments, inform our consideration of the recognition of variability as a feature in political and social thinking. We conclude by suggesting some approaches to management in the context of uncertainty.

CONCEPTUAL BASIS
Performance Measures

Students of systems behavior have recognized three characteristics by which the performance of a system may be evaluated: reliability, resilience, and vulnerability. All three relate to the concept of failure. We may expect that a sustainable system is one that can cope with failure in a noncatastrophic manner. Examples of these concepts can be found in many natural sciences, such as ecology and hydrology, and also in work on societal adaptation to hazards, including climatic change. We use definitions from hydrology (Hashimoto et al. 1982), supplemented by climatic adaptation research (Burton 1992; Smit 1993).

Reliability denotes how often a system is in a satisfactory state (Hashimoto et al. 1982). It is sometimes defined as the probability that no failures occur within some period of time. By satisfactory, we mean that the system operates to satisfy demands without crossing some meaningful threshold. For example, the system is able to provide some predetermined amount of water, such as the safe yield from a reservoir. Reliability may be assessed by examination of the record of consumption, counting how many times demand exceeded the threshold value of sustainable supply.

Resiliency describes the power of recovery of a system after failure. In hydrology, resiliency has been measured as the inverse of the expected length of time the system remains in an unsatisfactory state after a disturbance (Hashimoto et al. 1982). The concept of resiliency sometimes, as in most hydrological applications, implies "stability": the system returns to the preexisting equilibrium state after a disturbance or failure. This concept has also been termed "steadfastness" (Burton 1992). In social adaptation work, resiliency has been conceived of as "elasticity" (not to be confused with the economic concept). Here, elasticity recognizes that society effects an adaptive recovery, but not necessarily to the preexisting equilibrium. The concept has "a certain fuzziness," with respect to analytical evaluation, which serves to reinforce the idea that choices are "partly political and not solely technical" (Burton 1992, 266).

Vulnerability is a measure of the magnitude of failure, of the degree to which the system is adversely affected (Hashimoto et al. 1982; Smit 1993). Vulnerability is conditioned by the reliability and resiliency characteristics of the system; an unreliable or unresilient system may be prone to frequent or high-magnitude (or both) disruption. Vulnerability of the system itself may be recognized in impacts on components of the system, such as reservoir overflow in times of flood or deterioration of aquifer quality in times of drought. Social vulnerability to failure in the system may be recognized in allocation shortfalls and declining productivity in agriculture.

Uncertainty

Following Wynne (1992), uncertainty can be divided into four categories: (1) *risk*, when the odds are known; (2) *uncertainty*, when the odds are not known but the key variables and their parameters are known; (3) *ignorance*, when it is not known what should be known, and it is not even recognized which questions should be asked; and (4) *indeterminacy*, when causal chains or networks are open, and understanding is not possible even if appropriate questions are recognized and posed. In this chapter, our attention is focused upon the first two categories (risk and uncertainty). In other words, we do understand that variation will occur in the natural environment. However, we do not always know how much rain will fall, how long a drought will persist, or what the climate of the future may be.

THE CASE OF ISRAEL

Galnoor (1978) identified two periods in the development of water resources and associated policy in Israel; a third has occurred since he wrote his article. These can be seen in the time series of total water consumption (Figure 7.1).

Initially, the problem facing water managers was gaining access to supply. Surface water and wells had to be developed. Geographical imperatives suggested a nationally integrated system providing interregional transport and interseasonal storage was desirable: most of the water resource is in the north, while development of arable land is along the coast and in the more arid south; and, the Mediterranean climate is characterized by lack of rainfall in the summer. This initial phase culminated in the

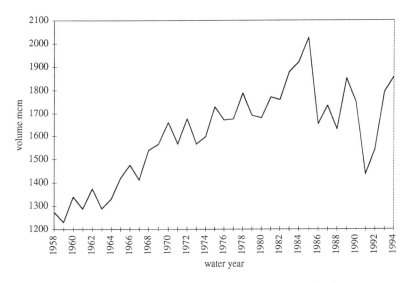

FIGURE 7.1. Total annual water consumption in Israel. Data are for the water year, autumn to autumn. Source: Water Allocation Department 1986; Sofer 1996.

opening in 1964 of the National Water Carrier, which moved water from north to south, using Kinneret as the major surface reservoir, and integrating many wells in the Coastal and Mountain Aquifers into a national system. Consumption grew by 150 percent from less than 1,400 million cubic meters per year to about 2,000 mcm/yr in the mid-1980s. Moreover, the growth rate of water consumption accelerated in the second period, compared to the first period. From 1958 to 1964, consumption grew at a rate of 11.3 mcm/yr, whereas from 1965 to 1985, the rate was 23.1 mcm/yr. From 1986 to 1994, the growth rate of 6 mcm/yr was overwhelmed by the considerable interannual variability.

According to Galnoor, the underlying principle for this first phase was *ideological rationality*. The social purposes of achieving self-sufficiency in agriculture and securing the land and its borders by settlement were paramount, regardless of economic cost. Water for agriculture was heavily subsidized by the government, which owned all water resources.

As Galnoor was writing, the growth rate of consumption was slowing somewhat from the initial steep increase in the first years after the National Water Carrier had opened. The amount of new water available to be added to the system was decreasing, as most renewable supply had been identified and tapped. Increases in supply were achieved, rather, by intensified efficiencies, reuse of treated wastewater, and drawdown of the aquifers. The water policy problem became one of scarcity, of making best use of available resources. Galnoor argued that ideological rationality was no longer an adequate or appropriate basis for water policy. Rather, management needed to shift to a basis in *economic rationality,* in which the price of water would reflect its true cost of production. Economic rationality would dictate best uses, and allocations would be adjusted accordingly.

The system continued to see growing consumption, with the rate of growth actually accelerating in the early 1980s. This growth in consumption was supported by increased mining of groundwater reserves. Saltwater intrusion into the Coastal Aquifer was experienced as a widespread problem, and the pending water crisis was being identified, even in the 1970s (Shuval 1980).

Since Galnoor wrote, an intense drought in the mid- to late 1980s brought about a period of crisis. Consumption fell by 20 percent or more and exhibited much greater interannual variability than in the preceding periods (Figure 7.1). In fact, the coefficient of variation (the ratio of standard deviation to mean) changed from about 300 percent for 1958–

1985 to about 3,300 percent for 1985–1994; in contrast, the coefficient of variability for total rainfall volume has been about 25 percent for 1932–1985 (calculated from data in Stanhill and Rapaport 1988). The problem facing water policy had become one of managing crisis. The overdraft of the aquifers had become equivalent to one year's total consumption and, together with salinization problems, could no longer be ignored. Across-the-board reductions in allocations and adjustments in sectoral shares were the result. Notably, agriculture's share of total consumption fell from about 70–75 percent to about 55–60 percent. This reduction in agricultural allocation, and moves toward emphasis on high-value crops, illustrate the scenario suggested by Allan (1995), in which agriculture becomes the reserve sector in order to protect municipal and industrial allocations. This argument is firmly rooted in an economic rationality basis for water management.

PERFORMANCE EVALUATION OF ISRAEL'S WATER SYSTEM

To assess reliability and the other performance indicators, a definition of "failure" is needed. Given the annual and national scales of analysis used here, a simple criterion for failure might be total consumption in excess of production (supply). Production here refers to water from surface flows and withdrawals from wells. Saline, reused, and recharged waters are included in this total, but amount to only about 10–11 percent of the total in the period 1959–1985 (Water Allocation Department 1986, Tables 29, 30). The ratio of total consumption to production exceeded 1.0 only in three years in the period 1958–1986 (Figure 7.2). (Comparable production data were not available at this writing for years after 1986.) These data may imply an average frequency of failure of one in ten. However, the annual ratio was always above 0.9 throughout the period of record (if net production—that is, minus saline, recharge, and reuse—is considered, the ratio is above 1.0 throughout the period). In the decade following the opening of the National Water Carrier, the margin between consumption and production was widest. From the mid-1970s, the ratio climbed toward 1.0. In the years immediately before the mid-1980s drought, the system was operating with very little margin—that is, close to failure (as it had been prior to the opening of the National Water Carrier).

Although failures were few, the reliability of the system was bought

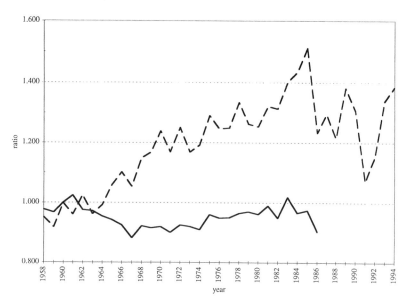

FIGURE 7.2. Definition of "failure" (ratios in excess of 1.0; see text) in the Israeli water system. Solid line, ratio of total annual consumption to annual production. Dashed line, ratio of total annual consumption to a threshold of 1,340 mcm, representing the average renewable supply due to rainfall. Source: Water Allocation Department 1986; Sofer 1996.

by mining of groundwater reserves, particularly in the Coastal Aquifer (Figure 7.3). The progressive fall of groundwater levels attests to this mining; the breaks in slope appear to match the phases of development of the system and policy identified above. The accumulated deficit in 1985 was equivalent to the annual supply of water (Schwarz 1994, 71). The exceptionally wet year 1991–1992 restored aquifer levels to where they had been decades earlier, but such recharge is not sustainable nor to be expected. Aquifer storage would not stay at that level unless groundwater exploitation were halted.

The definition of failure just discussed assumes that, in fact, supply (production) and demand (consumption) are independent. Ascription of cause and effect, even just specifying temporal precedence, is difficult. Consumption, it would seem, should be limited by supply, and supply should be limited by renewable endowment (precipitation) plus mining of reserves. Supply, therefore, should be the causal variable, and demand should be limited by, and adjusted to variability in, renewable supply;

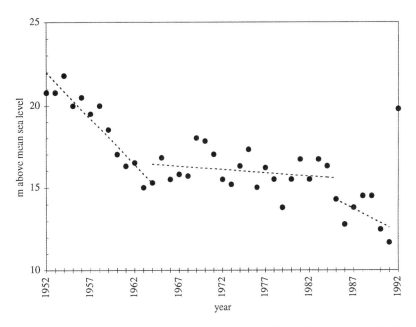

FIGURE 7.3. Annual high-water levels in a representative well in the Yarqon-Tani-nim Aquifer, Israel. Data interpolated from Figure 5 in Schwarz 1994. Linear trend lines fit by least-squares regression.

that would be an environmentally rational approach. In the Israeli situation, however, demand has been given precedence as the causal variable, and the water crisis has involved attempts to increase an assured supply.

A more sensitive index of failure, then, might be comparison of total consumption to annual replenishment by rainfall. Mean annual replenishment is about 1,340 mcm, varying between 600 and 2,370 mcm since the 1930s (Schwarz 1994, 69). By this measure, the system has been in failure almost continuously since the mid-1960s (cf. Figure 7.1). Only in the dry year of 1990–1991 did total consumption fall below this threshold; the extremely wet year of 1991–1992 prompted a rise in total consumption. Use of groundwater beyond annual replenishment amounts and reuse of treated wastewater have made the continued growth of consumption possible.

At no time in the period 1958–1986 did failure (by the first definition) occur in succeeding years. Recovery after failure was accomplished rapidly, with the ratio falling below 1.0 in the following year. Therefore, the system might appear to be resilient. The continued growth of consumption, even if slowed somewhat in the mid- to late 1970s, may be interpreted as an attempt to maintain stability. The mining of groundwater was important in this behavior. After the 1980s drought, there was an apparent adaptive recovery to a new equilibrium. Total consumption was reduced by 20 percent from the peak value of the mid-1980s, and the share of agriculture dropped from mid–70 percent to about 60 percent. The high variability that occurred after the mid-1980s may imply competing goals of recovery to preexisting equilibrium and adaptation to a new state. Climate also was certainly variable; after the severe drought, 1991–1992 was one of the wettest years of the century (and 1992–1993 was also wetter than average). Such variability may not have helped the attempt to adjust to a new reality of lower allocations.

The vulnerability of the system is best illustrated by the postdrought adjustment of consumption and allocations. Since 1985, annual total consumption fluctuated about a mean of 1,700 mcm, with large interannual variability; agriculture's share of total consumption fell from just over 70 percent to just over 60 percent, and although it has recovered somewhat in the mid-1990s, it is still below what might have been produced by the long-term trend of declining proportions. The system simply could not continue to be operated as it had, due to the accumulated debt to groundwater, growing demand in municipal and industrial sectors, and the moves toward peace. The practices that had placed the system in a precarious state with respect to reliability also made it vulnerable to "catastrophic" impact (in the sense that return to preexisting equilibrium is highly unlikely) of an extreme event.

The second-order impact on society is examined by considering the vulnerability of the agricultural sector to these changes in the water system. An agricultural production (quantity) index for crops was examined (Central Bureau of Statistics 1994, Table 13.1). The index exhibits a very strong positive trend over time. The interannual variability is isolated from the long-term trend by considering the change in the index from one year to the next. This variability may be due, in part, to variation in input of water. Prior to the mid-1980s, interannual change was always positive (except for one dry year) (Figure 7.4). The first dry year preceding the drought, 1983, had a large effect on production, as the index fell sharply. After the mid-1980s, there was very large variability in behavior of the index; this increased variability corresponded to that in total water consumption and in agriculture's share.

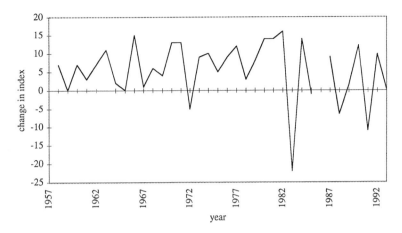

FIGURE 7.4. Interannual changes in the crop productivity (quantity) index for Israeli agriculture. Values represent the difference between one year's index and the preceding year's index. The base index was redefined in 1986, breaking the continuity of the record. Source: Central Bureau of Statistics 1994, Table 13.1.

WATER IN NATIONAL MANAGEMENT POLICIES

We have seen thus far that the water resource system in Israel has experienced a crisis of reliability, due to management that took its cue more from ideological rationality than from economic or environmental rationality. We now ask how management of water is viewed in the policies of the riparian parties. Are there signs that both economic and environmental considerations are the basis for planning for the severe stresses that the resource will experience in the coming years?

Israel

The summary report prepared for Israel's Year of the Environment explicitly recognized that "[s]hortage of water may be the most crucial environmental problem facing Israel today, touching upon its very existence" (Gabbay 1994, 17). The report went on to state that a water management plan needs to be based on: (1) maintenance of long-term reserves, (2) ability to cut allocations in time of drought, and (3) development of unconventional sources. Schwarz (1994, 71) stated that the new master plan of 1988 was premised on "the following data and forecasts: water demand was expected to grow with increasing population and standard of living; reliable future supply required restoration of groundwater stocks; intensive human activities in water catchments endangered the quality of water sources." A need for 2,000 mcm or more water per year was forecast for the next decade. The last time annual use reached that level (in the mid-1980s), it was not sustainable. Although agricultural allocations were reduced, and the contribution of agriculture to the national economy is small, there remain dreams for new agricultural projects. These include large-scale, intensive, high-value enterprises, such as olive and citrus groves and fish ponds in the northern Negev Desert (Rabinovich 1995–1996). Such projects will mostly use treated wastewater from the Tel Aviv conurbation (about 150–200 mcm/yr) or newly mined groundwater from under the Negev. Some crops, such as citrus, may require high-quality water and, given international competition, not return value from exports equivalent to their cost of production. Yet, these projects indicate the difficulty with which agricultural activity will be given up as a desirable national goal (perhaps harking back to ideological purposes), and the continuing pressure for water that will keep consumption very near to full utilization, if not beyond.

Jordan

Of all the parties involved in the Jordan River basin, Jordan is the one in most acute water shortage. The potential annual supply is in the range of 1,000–1,200 mcm, little more than half of Israel's peak use for a comparable population. As with Israel, most of Jordan's annual consumption is by agriculture (75 percent in 1993); the demand for irrigation is greater than the supply available. Total demand is expected to grow from under 900 mcm/yr in the early 1990s to reach 1,700 mcm/yr in the next decade (Salameh and Bannayan 1993; Shatanawi and Al-Jayousi 1995). Also, agriculture has served an important social-planning role in the country, as in Israel, and is still seen as a major source for employment in the future. Water policy calls for enhancing supplies (more dams and groundwater development), utilizing nonconventional sources (waste and brackish waters), improving efficiencies and conservation (by technology and management), and comprehensive planning to include reallocation among sectors (including limiting growth of allocation to agriculture) and among regions (Al-Fataftah and Abu-Taleb 1992). Still, most forecasts are for major deficits, based on growth trends as well as present and likely supplies.

Palestinian Authority

Total consumption per capita is lowest in the West Bank and Gaza areas; in 1990, the figure was 129 m³/yr, compared to 255 m³/yr in Jordan and 376 m³/yr in Israel (Awartani 1994). Projections suggest a doubling of water consumption in the next decade, due to growth of population, rising living standards, and expansion of agriculture and the amount of irrigation (Al-Khatib and Assaf 1994; Awartani 1994). Al-Khatib and Assaf (1994) expected per capita water demands would be at the level of Israeli demands by the year 2020. Almost all the demand is currently met by groundwater, and this situation is expected to continue. The situation implies the use of replenishable and fossil groundwater.

WATER IN THE PEACE INSTRUMENTS

Water, because it is a scarce shared resource, is recognized as a key point for negotiation and agreement between the contending parties. It has been proposed that water may exacerbate conflict or be used to foster peace (e.g., Gleick 1994; Lowi 1992; Baskin 1993; Kally 1993; Kliot 1994; Wolf 1995). The peace treaty between Israel and Jordan and the interim agreement between Israel and the Palestinians both recognize the scarcity of water now and the inexorable growth in demand. Do these instruments contain any shifts in policy-making that might illustrate that lessons have been learned about living with variability, scarcity, and uncertainty?

The Treaty of Peace between the State of Israel and the Hashemite Kingdom of Jordan was signed on 26 October 1994. Article 6, paragraph 3 states: "The Parties recognize that their water resources are not sufficient to meet their needs. More water should be supplied for their use through various methods, including projects of regional and international co-operation." Paragraph 4 identifies "ways to alleviate water shortage," which include: development of existing and new resources; increasing water availability (by regional projects); minimizing waste; prevention of contamination; mutual assistance; transfer of information; and joint research and development. Annex II, on Water Related Matters, details agreement for allocations of Jordan and Yarmuk Rivers water, and Arava groundwater. Israel promises, for example, that Jordan's allocation of Yarmuk water can increase by 50 mcm/yr; in part, this is accomplished by reducing Israel's withdrawal by 25 mcm/yr. Although this total amount is only about 3 percent of Israel's total consumption, it must be found from a system that already is at full utilization, and found consistently regardless of the interannual variation of supply and demand. The treaty emphasizes finding new water, in particular implying international projects. The underlying policy is to maintain (in the case of Jordan, increase) consumption rates. There is little room here for flexibility in the face of continued environmental variability.

The Israeli-Palestinian Interim Agreement on the West Bank and the Gaza Strip ("Oslo II") was signed on 28 September 1995. It is the first step in a process leading to Palestinian autonomy, which is envisioned to take many years. Water is only mentioned in the context that the Palestinian Authority's "Territorial jurisdiction includes land, subsoil and territorial waters." Issues of groundwater include historic, present, and future use and ownership, transterritorial extent of the aquifer, and the relative rights of the recharge and outlet riparians (e.g., Schiff 1995b). These issues are contentious, and the parties have agreed to defer their discussion and resolution to later in the peace process. For the time being, Israel has agreed to transfer approximately 28 mcm/yr to the Palestinians. Again, this amount must be found from a system that is at full utilization, and found consistently regardless of the interannual variation of supply and demand.

On 13 February 1996, Israel, Jordan, and the Palestinian Authority signed a Declaration of Principles for Cooperation on Water-related Matters and New and Additional Water Resources. The objectives for cooperation include: identifying needs for new and additional water; identifying and developing new and additional water sources; enhancing water supply and increasing efficiency of its use. Drought is specifically recognized as a threat: "The Core Parties will take appropriate measures in periods of drought and water scarcity." The approach implied here seems to be a response to hazard, rather than proactive planning. Among other water-related matters listed for cooperation are climatology, weather modification, and "sustainable water-related natural resource management." The document seems to emphasize the technological approach to water scarcity, including development and importation of new water. Yet there is also an opportunity for introducing concepts related to sustainability into policy-making and long-range planning. Such concepts are discussed later in the chapter.

Syria and Lebanon are the two riparians to the Jordan basin that have not yet negotiated peace with Israel. Water will be a critical component of negotiations with Syria, as the headwaters of the Jordan River are in part on the Golan Heights, now annexed by Israel. Many political observers feel that Lebanon will follow Syria's lead. Syria may attempt to gain riparian rights to the Kinneret, which Israel is likely to view as a security threat to its entire water economy (e.g., Schiff 1995a). Syria is also a riparian to the Yarmuk, using about 200 mcm/yr and certain to demand a large share in any settlement (Sofer 1996, personal communication). Israel would return to its former position as a minor riparian (if it were to leave the Golan Heights, it would give up even this position) if its arrangements with Syria and Jordan include the Yarmuk in a regional context. Unlike Israel, Jordan, and the Palestinian Authority, Syria and Lebanon are relatively water-rich. If negotiations emphasize water need rather than water rights (Baskin 1995), agreements on regional planning might proceed more quickly.

IMPLICATIONS

Environmental Rationality

The evidence presented above suggests that the co-riparians in the Jordan River basin are in a precarious position. Water is in short supply, with perhaps barely enough to meet current demand. Natural fluctuations and variability exacerbate this scarcity, because the resource has been developed and allocated for nearly full utilization. Growing populations, rising standards of living, expansion of demand in all sectors, and peace arrangements all generate demand for even more water. Plans for meeting this demand include technological approaches to using more resources within the entities with greater efficiency and to devising regional movement of water. Environmental variability is accorded little more than a nod of recognition. The performance evaluation of the Israeli system suggests a crisis of reliability in the system; living at full utilization, and borrowing supplies against the future, have left little margin for absorbing further shocks. Some positive signs are seen, in improved efficiencies (technological and managerial), peace, and talks for regional cooperation. Yet plans by all authorities are for more water and continued full utilization. The system will, therefore, likely remain unreliable and vulnerable.

Sustainable Water Management and Development

At the beginning of this chapter, we suggested that management to achieve water security should be pursued in the context of sustainable development (Koudstaal et al. 1992). A set of guidelines or principles for sustainable development should be formulated in the context of environmental variability and the Declaration of Principles for Cooperation on Water-related Matters and New and Additional Water Resources. Precedents exist for development of such principles or guidelines. The Dublin Statement, which emerged from the International Conference on Water and the Environment in Dublin in late January 1992, provided four principles which also reflect concern about the natural variability in water resource systems (Young et al. 1994, 161–162; see inset "Principles from the Dublin Conference"). A second precedent comes from Sustainability Principles for Water Management in Canada, prepared by the Canadian Water Resources Association (Mitchell and Shrubsole 1994, 5; see inset "Sustainability Principles from CWRA"). The "sustainability ethic" at the beginning of these principles for Canada explicitly recognizes the need to incorporate an environmental rationality into water management. The Water Science and Technology Board of the U.S. National Research Council is coordinating (as of 1996) a study on sustainable water resources for the Jordan River watershed, incorporating scientific and technological approaches to avoiding overexploitation, enhancing supply, and preserving environmental quality (David and Policansky 1996).

From our perspective, a valuable exercise would be to develop a comparable set of sustainable water management principles for the Jordan River states. Such principles could be based upon, but extend, the ideas contained in the Declaration of Principles for Cooperation on Water-related Matters and New and Additional Water Resources signed by Jordan, the Palestinians, and Israel. The declaration includes statements regarding both economic and environmental aspects of water. For example, it states that, regarding charges, water will not be supplied free of charge by any of the Core Parties. Furthermore, a section entitled "Principles of Cooperation" is explicit that "projects will be technically, economically and financially sustainable" and that "all Projects will be based on environmentally sound principles." However, the nature of such "environmentally sound principles" is not explained. Wolf and Murakami (1995) suggested a decision-making framework that included measures of technical, environmental, economic, and political viability. Our approach to environmental rationality, suggested here, contributes to their technical and environmental measures, in particular by focusing on the question of variability.

We recommend that one key principle should be that water allocation

Principle 1

FRESH WATER IS A finite and vulnerable resource, essential to sustain life, development, and the environment.

Since water sustains life, effective management of water resources demands a holistic approach, linking social and economic development with protection of natural ecosystems. Effective management links land and water uses across the whole of a catchment area or groundwater aquifer.

Principle 2

WATER DEVELOPMENT and management should be based on a participatory approach, involving users, planners, and policymakers at all levels.

The participatory approach involves raising awareness of the importance of water among policymakers and the general public. It means that decisions are taken at the lowest appropriate level, with full public consultation and involvement of users in the planning and implementation of water projects.

Principle 3

WOMEN PLAY A central part in the provision, management, and safeguarding of water.

The pivotal role of women as providers and users of water and guardians of the living environment has seldom been reflected in institutional arrangements for the development and management of water resources. Acceptance and implementation of this principle require positive policies to address women's specific needs and to equip and empower women to participate at all levels in water resources programs, including decision-making and implementation, in ways defined by them.

Principle 4

WATER HAS AN economic value in all its competing needs and should be recognized as an economic good.

Within this principle, it is vital to recognize the basic right of all human beings to have access to clean water and sanitation at an affordable price. Past failure to recognize this economic value of water has led to wasteful and environmentally damaging uses of the resource. Managing water as an economic good is an important way of achieving efficient and equitable use, and of encouraging conservation and protection of water resources.

CWRA SUSTAINABILITY PRINCIPLES FOR WATER
MANAGEMENT IN CANADA

Sustainability Ethic

WISE MANAGEMENT of water resources must be achieved by
genuine commitment to:

· ecological integrity and biological diversity to ensure a
  healthy environment;
· a dynamic economy; and
· social equity for present and future generations.

Water Management Principles

ACCEPTING THIS sustainability ethic, the CWRA will:

1. Practice integrated resource management by:
   · linking water quality and quantity and the management of
     other resources;
   · recognizing hydrological, ecological, social, and institu-
     tional systems; and
   · recognizing the importance of watershed and aquifer
     boundaries.
2. Encourage water conservation and the protection of water
   quality by:
   · recognizing the value and limits of water resources, and
     the cost of providing water in adequate quantity and
     quality;
   · acknowledging water's consumptive and nonconsumptive
     values to both humans and other species; and,
   · balancing education, market forces, and regulatory sys-
     tems to promote choice and recognition of the respon-
     sibility of beneficiaries to pay for the use of the resource.
3. Resolve water management issues by:
   · employing planning, monitoring, and research;
   · providing multidisciplinary information for decision-
     making;
   · encouraging active consultation and participation among
     all members of the public;
   · using negotiation and mediation to seek consensus; and
   · ensuring accountability through open communication,
     education, and public access to information.

decisions will be made with regard to natural variability in aquatic systems, and with regard to uncertainty of supplies due to such variability. Such a principle would be consistent with the first element of the sustainability ethic of the CWRA Sustainability Principles for Water Management in Canada, which stresses the need to consider ecological integrity and biological diversity.

We also recommend that another key principle should be that water allocation decisions will be made with regard to protecting the long-term integrity of aquatic systems. If this were to be included, then water planners and managers would be forced to address more explicitly the implications of "mining" groundwater aquifers as they seek to satisfy a growing demand for water in all of the cooperating parties.

In practice, then, explicit consideration of environmental variability, what Kay (1993) called an "environmental rationality," needs to be added to the economic rationality urged by many analysts. In his analysis, Kay (1993) showed that defining a threshold for maximum consumption linked to precipitation variability could have avoided most of the accumulated overdraft of groundwater in Israel. Benvenisti (1994) suggested a general mechanism for negotiations on water in the framework of peace agreements that illustrates this principle from another perspective. He suggested that equitable distribution should be agreed on the basis of guidelines for apportionment (for example, proportional shares), rather than on fixed amounts of water. Such an approach would offer greater flexibility for coping with variability than do present practices of specifying absolute amounts of water.

Complexity and Uncertainty

Recognition of the variability in natural systems highlights both the complexity and uncertainty with which planners and managers must deal. However, recognition of such variability is a first step in accepting that management always occurs in turbulent conditions, in which surprise and uncertainty are the norm rather than the exception. Such uncertainty poses special challenges for water managers, all the more so in a region, such as the Jordan River basin, in which high variability is the norm. For, as Christensen (1985, 63) has observed:

If people agree on what they want and how to achieve it, then certainty prevails and planning is rational application of knowledge. If they agree on what they want but do not know how to achieve it, then planning becomes a learning process; if they do not agree on what they want but do know how to achieve alternatives, then planning becomes a bargaining process; if they agree on neither means or ends, then planning becomes part of the search for order in chaos.

The comments by Christensen reinforce the interpretations by Wynne (1992) of risk and uncertainty, presented earlier in this chapter. That is, much of planning functions as if ends and means were known and agreed upon. In contrast, due to the considerable natural variability in environmental conditions, the reality is that often neither ends nor means are even known, let alone agreed upon. As a result, it is important that any approaches explicitly reflect the uncertainty and complexity associated with unpredictable natural variability.

One measure increasingly being invoked regarding complex and uncertain situations is the *precautionary principle*. As Principle 15 in the Rio Declaration states:

> In order to protect the environment, the precautionary approach shall be widely applied by States according to their capabilities. Where there are threats of serious or irreversible damage, lack of full scientific certainty shall not be used as a reason for postponing cost-effective measures to prevent environmental degradation. (United Nations Conference on Environment and Development 1992)

The precautionary principle stipulates that attention should be given to the environmental rationality of situations. However, while admonishing planners and managers to consider the environmental implications regarding alternative choices, the principle also is explicit that lack of complete understanding cannot be used as a justification for not taking action as long as measures to prevent environmental degradation are judged to be feasible.

Adaptive Management

If complexity, variability, and uncertainty are the norm, then it is unlikely that the *comprehensive rational,* or *synoptic,* planning model will be the most appropriate. As Mitchell (1989, 264) has noted, such models assume that a policy or management process moves through several phases:

1. A problem is clearly defined, and is separated from other problems of concern;

2. The goals, values, and objectives of the policymaker are identified. Furthermore, the policymaker is able to rank the goals, values, and objectives. In other words, *ends* are known;

3. The complete range of alternative solutions to the problem is identified. In other words, *means* for attaining the *ends* are known;

4. The consequences of each alternative relative to the problem, goals, values, and objectives are identified, compared, and ranked; and,

5. After comparing all of the alternative approaches or solutions, the policymaker selects the alternative that *maximizes* net expectations.

These activities characterize an ideal pattern of choice and decision-making, which rarely can be met. The main reason is that this approach assumes that the people involved are not only able to identify and rank goals, values, and objectives, but they also can choose consistently among them after having collected all the necessary data and systematically evaluated them. And, as noted already, all the necessary data are not available for the Jordan River system and region, and large natural variability poses significant problems in interpreting what averages mean for management purposes.

Given the weaknesses of the comprehensive rational planning model, and given the complexity and uncertainty surrounding water management issues in the Jordan River states, we believe that systematic consideration should be given to the merits of *adaptive management* (Holling 1978; Lee and Lawrence 1986; Walters 1986; Lee 1993; Rondinelli 1993). In Rondinelli's (1993, 7) words, "the argument for greater flexibility and innovation . . . rests in part on the observation that development policies are complex, uncertain, and require flexible and experimental methods of implementation." Adaptive management accepts that variability, surprise, and change are inevitable. As a result, management should be approached with the expectation that our knowledge and understanding are incomplete and imperfect, that mistakes will be made, and that adjustments will be needed.

In terms of performance evaluation, adaptive management suggests that resiliency be defined as elasticity rather than stability or steadfastness. In this manner, we also must build in some buffering capacity to reflect the reality of significant variability in natural systems, such as fluctuating rainfall and temperatures in the Jordan River states, which have profound implications for water balances. The spirit of adaptive management is that management is an exercise of social learning, and that we must design approaches so that there is capacity to modify and make changes, especially when the natural system behaves as "practical joker" (Holling 1978).

REFERENCES

Allan, J. A. 1995. "The Role of Drought in Determining the Reserve Water Sector in Israel." *Drought Network News* 7 (3): 21–23.
Awartani, H. 1994. "A Projection of the Demand for Water in the West Bank and Gaza Strip, 1992–2005." In J. Isaac and H. Shuval (eds.), *Water and Peace in the Middle East,* pp. 9–31. Amsterdam: Elsevier.
Baskin, G. 1995. "Suggestions for the Negotiations on Water." Internet document maintained at http://www.ipcri.org/. Jerusalem: Israel-Palestine Centre for Research and Information.
———, ed. 1993. *Water: Conflict or Cooperation.* Jerusalem: Israel/Palestine Center for Research and Information.
Benvenisti, E. 1994. "International Law and the Mountain Aquifer." In J. Isaac and H. Shuval (eds.), *Water and Peace in the Middle East,* pp. 229–238. Amsterdam: Elsevier.
Brooks, D. 1994. "Economics, Ecology and Equity: Lessons from the Energy Crisis in Managing Water Shared by Israelis and Palestinians." In J. Isaac and H. Shuval (eds.), *Water and Peace in the Middle East,* pp. 441–450. Amsterdam: Elsevier.
Burton, I. 1992. "Regions of Resilience: An Essay on Global Warming." In J. Schmandt and J. Clarkson (eds.), *The Regions and Global Warming,* pp. 257–274. New York: Oxford University Press.
Central Bureau of Statistics. 1994. *Statistical Abstract of Israel 1994, No. 45.* Jerusalem.
Christensen, K. S. 1985. "Coping with Uncertainty in Planning." *Journal of the American Planning Association* 51 (1): 63–73.
David, S., and D. Policanksy. 1996. "The Middle East: A Future Written in Water." *WSTB, Newsletter from the Water Science and Technology Board* 13 (4): 1–2.
Al-Fataftah, A-R., and M. F. Abu-Taleb. 1992. "Jordan's Water Action Plan." In E. J. Schiller (ed.), *Sustainable Water Resources Management in Arid Countries: Middle East and North Africa. Canadian Journal of Development Studies* (special issue): 153–171.
Fishelson, G. 1995. "Addressing the Problem of Water in the Middle East." In S. L. Spiegel and D. J. Pervin (eds.), *Practical Peacemaking in the Middle East.* Volume 2, *The Environment, Water, Refugees, and Economic Cooperation and Development,* pp. 117–137. New York: Garland Publishing Inc.
Gabbay, S. 1994. *The Environment in Israel.* Jerusalem: Ministry of the Environment.
Galnoor, I. 1978. "Water Policy Making in Israel." *Policy Analysis* 4 (3): 339–367.

Gleick, P. 1994. "Water, War & Peace in the Middle East." *Environment* 36 (3): 6–15, 35–42.

Hashimoto, T., J. R. Stedinger, and D. P. Loucks. 1982. "Reliability, Resiliency, and Vulnerability Criteria for Water Resource System Performance Evaluation." *Water Resources Research* 18 (1): 14–20.

Holling, C. S., ed. 1978. *Adaptive Environmental Assessment and Management.* Chichester, England: John Wiley.

Kally, E. 1993. *Water and Peace: Water Resources and the Arab-Israeli Peace Process.* Westport, Conn.: Praeger.

Kay, P. A. 1993. "Recognition of Climatic Sensitivity as an Element in the Management of Shared Water Resources." In *Proceedings of the International Symposium on Water Resources in the Middle East: Policy and Institutional Aspects,* pp. 108–113. Urbana, Ill.: International Water Resources Association.

Al-Khatib, N., and K. A. Assaf. 1994. "Palestine Water Supplies and Demands." In J. Isaac and H. Shuval (eds.), *Water and Peace in the Middle East,* pp. 55–68. Amsterdam: Elsevier.

Kliot, N. 1994. *Water Resources and Conflict in the Middle East.* New York: Routledge.

Koudstaal, R., F. R. Rijsberman, and H. Savenije. 1992. "Water and Sustainable Development." *Natural Resources Forum* 16 (4): 277–290.

Lee, K. N. 1993. *Compass and Gyroscope: Integrating Science and Politics for the Environment.* Washington, D.C.: Island Press.

———, and J. Lawrence. 1986. "Restoration under the Northwest Power Act." *Environmental Law* 16 (3): 431–460.

Lonergan, S. C., and D. B. Brooks. 1994. *Watershed: The Role of Fresh Water in the Israeli-Palestinian Conflict.* Ottawa: International Development Research Centre.

Lowi, M. R. 1992. "West Bank Water Resources and the Resolution of Conflict in the Middle East." In *Occasional Paper Series of the Project on Environmental Change and Acute Conflict,* pp. 29–60. Toronto: University of Toronto.

Mitchell, B. 1989. *Geography and Resource Analysis,* 2d ed. Harlow, England: Longman Scientific and Technical.

———, and D. Shrubsole. 1994. *Canadian Water Management: Visions for Sustainability.* Cambridge, Ont.: Canadian Water Resources Association.

Rabinovich, A. 1995–1996. "New Lease on Life." *JNF Illustrated* Winter: 4–8. Jerusalem: Jewish National Fund.

Rondinelli, D. A. 1993. *Development Projects as Policy Experiments: An Adaptive Approach to Development Administration,* 2d ed. New York: Routledge.

Salameh, E., and H. Bannayan. 1993. *Water Resources of Jordan: Present Status and Future Potentials.* Amman, Jordan: Friedrich Ebert Stiftung.

Schiff, Z. 1995a. "They Are Forgetting the Golan's Water." *Ha'aretz,* June 7.

———. 1995b. "Again Forgetting the Water." *Ha'aretz,* July 11.

Schwarz, J. 1994. "Management of Israel's Water Resources." In J. Isaac and H. Shuval (eds.), *Water and Peace in the Middle East,* pp. 69–82. Amsterdam: Elsevier.

Shatanawi, M. R., and O. Al-Jayousi. 1995. "Evaluating Market-Oriented

Water Policies in Jordan: A Comparative Study." *Water International* 20 (2): 88–97.

Shuval, H. I. 1980. *Water Quality Management under Conditions of Scarcity.* New York: Academic Press.

Smit, B., ed. 1993. *Adaptation to Climatic Variability and Change.* Report of the Task Force on Climate Adaptation, the Canadian Climate Program. Department of Geography Occasional Paper No. 19. Guelph, Ont.: University of Guelph.

Sofer, A. 1996. Personal communication (water-year consumption data for 1985–1994, from Water Commissioner). Department of Geography, University of Haifa.

Stanhill, G., and C. Rapaport. 1988. "Temporal and Spatial Variation in the Volume of Rain Falling Annually in Israel." *Israel Journal of Earth Science* 37 (4): 211–221.

United Nations Conference on Environment and Development. 1992. *The Rio Declaration on Environment and Development.* Geneva: UNCED Secretariat.

Walters, C. 1986. *Adaptive Management of Renewable Resources.* New York: Macmillan.

Water Allocation Department. 1986. *Water in Israel, Consumption and Extraction 1962–1985.* Tel Aviv: Water Commission, Ministry of Agriculture.

Wolf, A. 1995. *Hydropolitics along the Jordan River: Scarce Water and Its Impact on the Arab-Israeli Conflict.* Tokyo: United Nations University Press.

———, and M. Murakami. 1995. "Techno-political Decision Making for Water Resources Development: The Jordan River Watershed." *Water Resources Development* 11 (2): 147–162.

World Commission on Environment and Development. 1987. *Our Common Future.* Oxford: Oxford University Press.

Wynne, B. 1992. "Uncertainty and Environmental Learning: Reconceiving Science and Policy in the Preventative Paradigm." *Global Environmental Change* 2 (2): 111–127.

Young, G. J., J. C. I. Dooge, and J. C. Rodda. 1994. *Global Water Resource Issues.* Cambridge, England: Cambridge University Press.

8. A Cooperative Framework for Sharing Scarce Water Resources
ISRAEL, JORDAN, AND THE PALESTINIAN AUTHORITY

Nurit N. Kliot

INTRODUCTION

The purpose of this chapter is to present a framework for sharing the scarce water resources of Israel, Jordan, and the Palestinians, with the possible future affiliation of Syria and Lebanon—the other co-riparians. At the time of the initial writing of this chapter (summer of 1996) two major events were shaping the prospects for cooperative venture between Israel and its neighbors in the water sector: on the one hand, the peace treaty with Jordan and the Oslo II Agreement with the Palestinians paved the way legally toward mutual recognition of water rights and possibilities for cooperation on water issues. On the other hand, the victory of the nationalistic bloc in the 1996 elections threatened to slow down the peace process and even bring it to a halt altogether, thereby having immediate repercussions on the evolving patterns of cooperation between Israel and Jordan, and Israel and the Palestinians. Events in early 1998 confirmed this observation.

This chapter will be divided into two sections: in the first section we shall start by presenting the shared water resources of the Levant, and will then proceed with a survey of past proposals and plans for sharing water resources and the obstacles to their implementation; this part will conclude with detailed documentation of the stipulations concerning water in both the Israeli-Jordanian Peace Treaty and the Israeli-Palestinian Agreement on Water and Sewage. In the second part of this chapter, various forms of sharing water resources, and the principles which guide them, will be explored as possible frameworks for regional cooperation in water issues. Finally, a proposal for cooperative frameworks for the management of common water resources will be offered.

PART I: BACKGROUND

The Common Water Resources

The Jordan River and its tributary, the Yarmuk, are shared by the states of Lebanon, Syria, Israel, and Jordan and by the Palestinian Authority. The Upper Jordan (up to its southern exit from Lake Tiberias) has three sources: the Hasbani (Lebanon), the Banias (until 1967 in Syria; since 1967 under Israeli control), and the Dan, which was and is within Israeli territory. The Palestinians do not have any share in the Upper Jordan. The drainage basin of the Upper Jordan is shared by Syria, Israel, and Lebanon (Kliot 1994). The total flow of the Upper Jordan, some 540 mcm, is delivered to Lake Tiberias, and Israel utilizes all this water, mainly through its National Water Carrier, which pumps, on the average, 450–500 mcm of water per year (Kliot 1994; Wolf 1995a; Bakour and Kolars 1993, 131; Lonergan and Brooks 1993, 29). The Lower Jordan, from the southern tip of Lake Tiberias up to the Dead Sea, is shared by Syria, Israel, Jordan, and the Palestinians. The Palestinians claim riparian rights in the Jordan, as the "West Bank has about 90 km frontage on the banks of the Lower Jordan River" (Abu-Sway, Al-Jamal et al. 1994, 7). Had the water diversions included in the Johnston negotiations been developed, water from the Jordan system of some 70–150 mcm/yr would reach the Jordan Valley (Wolf 1995b, 141). A quota of 100 mcm within Jordan's allocation in the Johnston Plan was assigned for the West Bank's needs.

The Yarmuk, the most important tributary of the Jordan, has a discharge of 400 mcm (Salameh 1992, 99), 450 mcm (Kliot 1994, 184), or 500 mcm (Al-Weshah 1992; Bakour and Kolars 1993, 131). There is no argument on the patterns of utilization of the Yarmuk. Syria uses 190–200 mcm, Jordan 120–130 mcm (by the East Ghor Canal, or as it is called today, the King Abdullah Canal) (Al-Weshah 1992, 127; Hof 1995, 48). In the last decade, Israel has been using about 70–100 mcm a year (Hof 1995, 48; Kliot 1994). Israel has claimed riparian rights to the water of the Yarmuk since the 1970s.

As a result of the overutilization of the Jordan-Yarmuk system, the total discharge of the Jordan into the Dead Sea declined from 1,370 mcm/year to 250–300 mcm/year, mostly irrigation return flow, inter-catchment runoff, or saline spring discharges and sewage dumped by Israel into the Lower Jordan (Salameh 1992; Kliot 1994). The other shared water resources in the Levant are the groundwater resources shared by Israel and the Palestinians. These resources include two major systems of aquifers: the mountain aquifer and the coastal aquifer (see

TABLE 8.1. Sources and Uses of the Jordan-Yarmuk and Shared Aquifers (Before the Agreements on Water Sharing)

Sources	Volume (CCM)	Utilization (MCM)				Flow to the Dead Sea
		Syria	Jordan	Palestinian Authority	Israel	
Yarmuk	400–500	190–200	120–130		70–100	250–300 MCM/year (return irrigation water flow, inter-catchment runoff, saline springs and sewage)
Jordan	540	None	None		450–500	
Coastal Aquifer (Gaza)	60	Syria and Jordan are not riparians to groundwater resources		90–100*	3	
Mountain Aquifer	679			118	483	

*Overexploitation of this aquifer.

SOURCES FOR THE TABLE: For the Yarmuk: Salameh 1992: 184; Al-Weshah 1992; Bakour and Kolars 1994: 131.
  Utilization: Al-Weshah 1992: 127; Hof 1995: 45.
  Mountain Aquifer: Kahana 1994: 9; Kahan 1983; Abu-Sway 1994: 5.

Figure 8.1). The mountain aquifer is divided into three subaquifer systems: the western aquifer (with a safe yield of 300–335 mcm/year), the northeastern aquifer (with 130–150 mcm safe yield), and the eastern aquifer (with 150–250 mcm safe yield). The total annual recharge of the mountain aquifer is 679 mcm (Kahana 1994, 9; Kahan 1987; Abu-Sway, Al-Jamal et al. 1994, 5).

The coastal aquifer in the Gaza Strip has an annual replenishment of around 60 mcm, but it is overexploited by about 30–50 mcm/year (total pumping of 90–110 mcm) (Nassereddin 1994, 64; Abu-Sway, Al-Jamal et al. 1994, 6). Present Palestinian extraction of groundwater resources in the West Bank is 118 mcm/year (according to Article 40 of the Water and Sewage Agreement), or ranging between 110 and 183 mcm/year according to Palestinian sources.

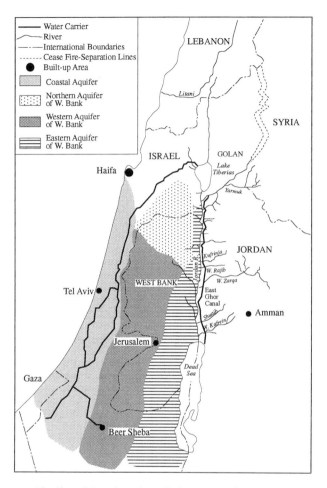

FIGURE 8.1. The Shared Aquifers of Israel, the West Bank, and Gaza.

Israel utilizes 483 mcm/year of the mountain aquifer (40 mcm from the eastern aquifer, 103 mcm from the northeastern aquifer, and 340 mcm from the western aquifer). Thus, Israel is currently using 63 percent of the mountain aquifer. Israel also pumps 3–4 mcm from the Gaza aquifer, but the 2–3 mcm year which Mekorot delivers is a net addition to Gaza (Lonergan and Brooks 1994, 134).

Jordan and Israel run their water economies at an annual deficit which is estimated at 220 mcm for Israel and 75–180 mcm for Jordan (Naff

1992, 108). The Palestinians, however, suffer from chronic water short-ages, and water supply to them does not fulfill international standards of 100 liters per person per day, generally regarded as the minimum neces-sary for human health and hygiene. Gaza's Palestinians certainly do not receive this amount of water (Lindholm 1992, 76–77). Thus, any insti-tutional framework for regional cooperation will have to find ways to supplement the present water resources. Also, it will have to determine how to distribute the meager water resources in a more equitable manner and manage the common water resources in the most efficient way.

Patterns of Cooperation, Obstacles to Cooperation, and Conflict over the Jordan-Yarmuk Resources

The Jordan-Yarmuk system was the target of no less than sixteen detailed plans for management and utilization of that basin (Kliot 1994, 187–198). Because of the Jewish-Arab conflict, many of the plans were not multipurpose integrated plans for the whole basin, and many tended to favor either Israel or the Arab states. Eventually, in the absence of any agreement over water sharing, Israel and the Arab states, most particu-larly Jordan, proceeded with their separate and often conflicting devel-opment plans. The alignment of the Israeli National Water Carrier had to change from its original location at the Jordan River, in the demilita-rized zone, to Lake Tiberias as the source (Lindholm 1992, 57; Wolf 1995a). The United States intervened in the tense situation (shooting skirmishes in 1953) and sent special envoy Eric Johnston to try and me-diate among the adversaries (see chapter by Hof in this volume). In 1964, the Arab states agreed on a diversion of the Upper Jordan tributaries to Arab territory, which would thus render the Israeli National Water Carrier (which was ready by then) useless. This decision led to violent border incidents in 1965 when Israel attacked the diversion works in Syria (Stevens 1965; Inbar and Maos 1984; Wolf 1995a). The 1967 Arab-Israeli War put an end to Arab control of the headwaters of the Jordan, and to their ability to divert these resources. Syria progressed in its development of the Yarmuk; Jordan completed the 110.5-km-long King Abdullah Canal, which delivers the Yarmuk water for irrigation in the Jordan Rift Valley; and Israel has taken full advantage of the Na-tional Water Carrier, which delivers the Jordan's water, pumped from Lake Tiberias, to agricultural areas in southern Israel.

An agreement to share the water resources of the Yarmuk was signed between Syria and Jordan in 1953. This agreement called for the construction of a storage dam (Unity Dam, Al-Wahda, or Maqarin Dam) on the Yarmuk River and the subsequent division of the water resources at a ratio of about 1:3 between Syria and Jordan, respectively (Salameh 1992). According to the 1987 version of the plan to construct that dam, it was envisaged to store 180–200 mcm/year (instead of 350 mcm/year, as in the original plan) (Taubenblatt 1988, 47; Salameh 1992). Now, as Syria is already utilizing half of the available water of the Yarmuk, it does not seem that the planned dam has any economic value. Israel has objected to the construction of a dam on the Yarmuk if its riparian rights to a quota of the Yarmuk's water are not acknowledged.

Since the mid-1980s, comprehensive studies on the strategic aspects of water in the Middle East have emerged, some with specific stipulations for basinwide cooperation. Kally, in a remarkable document published in 1986, called for regional projects which would bring fresh water to the water-scarce Middle East (Kally 1986). Kally called for importation of water from the Nile to the Sinai Peninsula, and to the Israeli dry south. He also called for importation of the Litani River's water to Israel, the West Bank, and Jordan. Although selling the Litani's water for a good price would be good business for Lebanon, Amery (1993) quite correctly showed that Lebanon will need the water of the Litani for the development of the economy of southern Lebanon.

Another wave of proposals for regional cooperation in water issues appeared in connection with the peace process initiated in Madrid in 1991. The spirit of peace encouraged many specialists to speculate on and even propose large-scale cooperative and joint projects in the water sector. The bases of most of the projects suggested were inclusion of the region at large, water importation from outside the region, and desalinization plants. All these projects were capital-intensive, and many would produce "new" water too expensive for most of the poor water-scarce countries of the Levant. The most prominent project was the importation of water from Turkey to the Middle East via a Peace Water Pipe Line, suggested in 1986 by Turgut Özal, the Turkish Prime Minister (Gruen 1993, 9; Kliot 1994). Various plans for desalination of either seawater or saline springs were proposed as cooperative projects among Israel, the Palestinians, and Jordan. The most important seawater desalination plant was suggested for Gaza either as an independent project or as a joint venture of Israel and the Palestinian Authority at Gaza (Shuval 1992).

In October of 1994 as part of the preparation for the Middle East and

North Africa Economic Summit at Casablanca (30 October 1994), the World Bank, Israel, Jordan, and Egypt submitted their proposals for regional economic cooperation. Important cooperative ventures were offered for the water sector—the most important being the Red Sea–Dead Sea Canal and a Med Sea–Dead Sea Canal. The Red-Dead Sea Canal was endorsed by both Israel and Jordan and was supported by the World Bank (World Bank 1994, 13). Israel also proposed three alignments for the Mediterranean-Dead Sea Canal—all based on a combination of cheap energy for desalination purposes (Government of Israel 1994, IV-6.1-18). This type of project attracted many researchers who dealt, in depth, with the engineering, technical, and economic aspects of such schemes and found them feasible from both the technical and economic perspectives (Murakami et al. 1995, 191; Murakami and Musaike 1995, 136–145). However, as the Rift Valley is environmentally extremely vulnerable, mainly because of its seismic activity, arid climate, and the fossil groundwater resources which could be affected by the project, the realization of these plans is doubtful. Another major obstacle is the high cost estimated for the Red-Dead Sea Canal. Most recent estimates (1995) put it at $5 billion, an investment perhaps too high for potential foreign investors. Finally, and most important, all the above proposed regional projects lacked the appropriate institutional structure for their functioning. As the subject of cooperation is a scarce resource and the riparians are already in deep conflict over its allocation and use, an institutional framework which will secure riparian rights and duties is almost mandatory.

The Israeli-Jordanian Peace Treaty and the Israeli-Palestinian Agreement on Water and Sewage of 1995

The foundations for an institutional framework for sharing water resources were laid in two different documents: first, the Treaty of Peace between Israel and Jordan refers to the Jordan-Yarmuk and to groundwater resources in the Wadi Araba-Arava, on their common border in the southern Rift Valley. The other document is Article 40 in the 1995 Israeli-Palestinian agreement, which deals with water and sewage. These two documents will be presented in detail, as they reflect some of the institutional principles advocated by the riparians and also principles which they would not like to abide by.

The Treaty of Peace signed between Israel and Jordan on 26 October 1994 (the Israeli version in English), in Article 6 and Annex II, addressed matters related to water resources shared by the two countries. In Article 6.1 of the treaty, the parties agree mutually to recognize the rightful allocations to both of them of the Jordan and Yarmuk River waters and also the Wadi Araba-Arava groundwater, in accordance with the agreed-upon acceptable principles as set out in Annex II (details below). In Articles 6.2 and 6.3, the parties acknowledge that common water resources are not sufficient to satisfy their needs, and agree on cooperation in order "to ensure that the management of their water sources does not harm the water resources of the other Party" (Government of Israel 1994, 7). The parties also agree that more water should be supplied for their use through various methods, including projects of regional and international cooperation, including the possibility of transboundary water transfers. Article 6.4 stresses the need for minimizing wastage of water resources, preventing contamination of water resources, mutual assistance in the alleviation of water shortages, the transfer of information, and joint research and development in water-related subjects (Government of Israel 1994, 7–8). The most important stipulations concerning common water resources are included in Annex II of the treaty. These provisions are summarized in Table 8.2. Beyond the quantities of water allocated to Israel and Jordan from specific sources and in specific periods, the peace treaty has other provisions which refer to measures to increase water quantities in the future and measures to preserve water quality.

First, the treaty has a provision to construct a diversion weir near Adasiyah that will enhance Jordan's water diversion to the King Abdullah Canal (Figure 8.1). Jordan would like to construct a reservoir which will hold more than 3 mcm. Israel and Jordan estimate the cost of construction at $10 million, and Jordan is currently in the process of contracting a foreign company to build that reservoir and weir—with Israel's consent.

Second, the two parties agreed on the rehabilitation and desalinization of the Lower Jordan. Israel is diverting saline springs from Lake Tiberias and sewage to the Lower Jordan. Israel and Jordan will seek international financial aid to cover the expenses of cleaning up the Lower Jordan. Jordan will be allocated 10 mcm out of the 20 mcm which flows in the Lower Jordan.

Third, Israel and Jordan will cooperate in the construction of small storage banks on the Jordan River, which will mostly be utilized by Jordan, for storing winter floodwater.

Fourth, the two parties have committed themselves to protecting both rivers from pollution, contamination, or unauthorized withdrawals from each other's allocations.

Fifth, a Joint Water Committee was established. The Joint Water Committee will be composed of three members from each country. The committee will specify its work procedures, the frequency of its meetings, and the details of its scope of work. The committee may invite experts and/or advisors as required, and can also establish a number of specialized subcommittees and assign them technical tasks. Basically, the purpose of the committee is to fulfill the provisions of Annex II.

A unique arrangement in Annex II guarantees that the fourteen wells (7 mcm/year) Israel uses in the Wadi Araba-Arava will continue to serve Israeli settlements, though the sovereignty over this territory was transferred to Jordan as part of the peace treaty. Israel is allowed to expand its water use of this groundwater by an additional 10 mcm/year, but as of this writing, Israel has not, in practice, been allowed to drill new wells for its expanded use. Apart from this, Israel and Jordan have respected all their other obligations under the treaty and Annex II.

Some unique characteristics of the above provisions should be highlighted. There is a great emphasis on mutualism—namely, that both Israel and Jordan do not make any concessions to the other party without receiving some in exchange. Also, there is a clear-cut attempt to preserve the current patterns of use of both parties. However, concessions have been made to Jordan, which cannot store and deliver winter flows. Israel stores and delivers this much-needed water to Jordan during the season of highest demand—summer. It is also interesting that not all the storage and projects which are planned for the future are joint ventures; there is only the decision to divide the benefits of new projects either equally or in Jordan's favor. There is no doubt that for the near future Jordan has ameliorated its severe shortage, and, in that respect, the peace treaty water provisions could be classified as "equitable" (see Part 2 of this chapter). The provisions, however, also point to what the treaty lacks: there is no reference to the other riparians, Syria, Lebanon, and the Palestinians. As Syria, as an upper riparian, unilaterally expands its water use of the Yarmuk, Israel and Jordan, the downstream riparians, may find that all treaty provisions concerning the Yarmuk are worthless. Thus, though the peace treaty legally acknowledges Jordan's and Israel's rights in the common water resources by ignoring the other riparians' rights, it may become worthless in the future.

TABLE 8.2. Water Allocation to Jordan and Israel According to Seasons and Sources According to The Treaty of Peace

Water Source and Season	Hashemite Kingdom of Jordan	Israel	Notes
Yarmuk River Winter Period 16.10–14.5	The rest of the flow[1]	13 MCM + 20 MCM[1]	According to Jordanian sources,[2] a net gain for Jordan is 37 MCM/year. Israel used to pump 50 MCM during the winter and deliver it to Lake Tiberias.
Summer period 15.5–15.10	"The rest of the flow" means 8 MCM	12 MCM	Israel used to pump 20 MCM.
Total[3]	130 MCM	25 MCM/year	Israel used to pump a total of 70 MCM and according to the Treaty limited itself to 25 MCM/year.
The Jordan River Winter 16.10–14.5	Jordan receives 20 MCM from the Jordan River upstream from Degania. This is a swap for the 20 MCM of winter flow which Israel is allowed to pump during the winter. Also Jordan is allowed to store 20 MCM of floods in the Lower Jordan and is entitled to 10 MCM of the desalinated water of the Lower Jordan.	Israel may use up to 3 MCM/year of added stored water and 10 MCM of desalinated water in the Lower Jordan. Israel may increase the current use of Wadi Arava groundwater by 10 MCM.	This allocation is based on Israel's capacity to store winter floodwater in Lake Tiberias and transfer it to Jordan, which lacks water storage facilities; the water is provided in the season of need—the summer. In the longer range, investment is needed for a desalination plant in the Lower Jordan. The water will be divided equally. Storage for floods in the Lower Jordan (7 MCM for Jordan, 3 MCM to Israel) is also a long-range project. Until the desalination plant is built, Israel will provide Jordan with 10 MCM upstream from Degania during the winter. For that Israel will be permitted to increase its use of the groundwater in the Arava—a "swap" for the 20 MCM it provides Jordan.

Total gain Short-range	50 MCM (currently 30 MCM)	30 MCM reduction in the short run	Increase in the water supply in the future depends on investments in dams, storage, and desalination facilities.
Long-range	Up to 100 MCM. From storage at the Yarmuk (20 MCM) and on the Jordan River (20 MCM) + 10 MCM desalinated water (50 MCM)	23–30 MCM increase in the long run	

[1] This amount is compensated for by the 20 MCM which Jordan receives from Israel.

[2] Dr. Munthar Haddadin, Jordan's representative in the water negotiations and S. Irsheidat, Jordan Irrigation Authority.

[3] Syria's withdrawal: 200 MCM, Israel 50–70, Jordan, 130, and 44 MCM flow to the Dead Sea.

SOURCES FOR THE TABLE: *Treaty of Peace*, 256.10.94; Hof 1995; *Jordan Times* 20–21 Oct. 1994; 1995.

On 18 September 1995, Israel and the Palestinians signed an Agreement on Water and Sewage, an important step toward alleviating and perhaps solving the conflict over their shared water resources. The agreement will be valid for the interim period as defined in the Oslo II agreement. In Article 40 of the 1995 agreement, Israel "recognizes the Palestinian water rights in the West Bank. These will be negotiated in the permanent status negotiations and settled in the Permanent Status Agreement relating to the various water resources" (Article 40, Water and Sewage, Israeli-Palestinian Agreement). The parties recognize the necessity to develop additional water for various uses (similar to the provisions in the peace treaty with Jordan). The parties agree to coordinate the management of water and sewage resources and systems in the West Bank during the interim period in accordance with the following principles:

1.   Maintaining existing quantities of utilization from resources;

2.   Preventing the deterioration of water quality;

3.   Ensuring sustainable use in the future, in quantity and quality;

4.   Adjusting the utilization of the resources according to climate and hydrology;

5.   Preventing any harm to water resources; and

6.   Treating, reusing, or properly disposing of all forms of sewage.

The other stipulations relate to the transfer of responsibilities and powers in the sphere of water and sewage in the West Bank in the areas in which the Palestinians assumed responsibility. An important part in the agreement is played by "additional water." Both sides have agreed that the future needs of the Palestinians in the West Bank are estimated as being between 70 and 80 mcm/year. In order to meet the immediate needs of the Palestinians for fresh water for domestic use, both sides recognize the necessity to make available to the Palestinians a total quantity of 28.6 mcm/year during the interim period. Israel is committed to adding 4.5 mcm to Hebron, Bethlehem, Ramallah, Salfit, Nablus, and Jenin, and also 5 mcm/year to the Gaza Strip. The Palestinians will have to provide 2.1 mcm to Nablus and 17 mcm/year from the eastern aquifer to Hebron, Bethlehem, and Ramallah. The rest of the water, 41.4–51.4 mcm/year, will be developed by the Palestinians from the eastern aquifer and other agreed resources in the West Bank. (Thus, the total of 70–80 mcm will be provided by an "immediate" 28.6 mcm and an additional 41.4–51.4 mcm/year which will be developed in the longer range.)

A Joint Water Committee (JWC) will be established in order to implement all the undertakings in the above agreement. The functions of the JWC were specified as coordinating management of water resources and water and sewage systems and protecting these systems; exchange of information; overseeing the operation of the joint supervision and enforcement mechanisms; and resolution of water- and sewage-related disputes. The Joint Water Committee will also be responsible for arrangements for water supply from one side to the other, for monitoring systems, and for other issues of mutual interest. As for structure, the JWC will be composed of an equal number of representatives from each side, and all its decisions will be reached by consensus. An immediate result of the agreement was the establishment of a Joint Supervision and Enforcement Team to supervise the enforcement of the agreement.

In the agreement, there are also stipulations for mutual water purchases, mutual cooperation concerning regional development programs, exchange of data reports, plans, etc., especially data on existing extractions. As for the Gaza Strip, Israel agreed that the Palestinians will assume responsibility for all the water and sewage installations except for those which serve the Israeli settlements and military installations. The two sides will establish subcommittees to deal with all issues of mutual interest.

Finally, the Israelis and Palestinians have agreed on the safe yield of their shared groundwater:

- *Western aquifer:* 362 mcm (22 mcm of it is used by the Palestinians).
- *Eastern aquifer:* 172 mcm (40 mcm used by the Israelis, 54 mcm by Palestinians).
- *Northeastern aquifer:* 145 mcm, 103 mcm of which is used by Israel and 42 mcm by the Palestinians.
  Total: 679 mcm/year.

The above agreement reveals the following principles:

1. There is an Israeli acknowledgment of Palestinian water rights to the West Bank aquifer.

2. Quantitative measures for Palestinian water rights in the West Bank were specified at 70–80 mcm; the agreement nearly doubled their present water quota for domestic use for the interim period, from the current 31 mcm/year to 59.6 mcm/year.

3. The achievement of the agreement is in its practical arrangements and pragmatism. Israel agreed to parity in managing the shared water resources and sewage in order to secure a responsible process of management. That side of the agreement has already proved its worth—a Joint Israeli-Palestinian Supervision and Enforcement Team was able to find and obstruct twelve illegal water drillings in the Jenin area. The experience gained in the joint teams and in the work of the Joint Water Committee was facilitating mutual confidence-building between Israel and the Palestinians. However, the cooperation within the Joint Water Committee was short-lived, as the relations between the Israelis and Palestinans deteriorated, mainly in 1997. The change of government in Israel, where there is very little goodwill toward Palestinian needs, almost completely paralyzed the work of the JWC.

4. Whereas joint management is practiced for the West Bank, Gaza is fully independent, except for the small number of wells which serve Israeli settlements in the Qatif area.

5. There is an emphasis on sewage treatment as a means both to prevent pollution and increase water supply to agriculture.

6. For the Palestinian Authority, the agreement enables the definition of water as a "public good" which does not belong to the landowners. If the Palestinian Authority assumes full responsibility, it will have to manage the quality and quantity of water resources and the number and capacity of wells, and assume authority over prevention of pollution. The implementation task is enormous: the water situation in both Gaza and the West Bank is severe. Israeli radio and newspapers are reporting a shortage of drinking water in Hebron at the time of the writing of this chapter. About 14 mcm of the water provided to Hebron is stolen by local residents, and large quantities are leaking from the deteriorated pipe system.

7. Finally, in February of 1996, Israel, Jordan, and the Palestinians signed a common Declaration of Principles for Joint Development of Water Resources. The declaration does not deal with allocation of present water resources but focuses on the development of new water resources. On the institutional side, the declaration calls for full coordination among water institutions and water laws of the three parties, importation of water from outside the region, desalination, cloud seeding, and so on. Again, there is avoidance of al-

locating shared water resources common to all. The Palestinians, Syria, and Lebanon are not part of the Jordan-Yarmuk according to the above two agreements.

PART 2: INSTITUTIONAL FRAMEWORK FOR SHARED WATER RESOURCES

Water Resource Doctrines

Management (and nonmanagement) and policies toward common water resources have been developing for hundreds of years, during which, basically, seven doctrines or theories concerning shared water resources have evolved.

The Doctrine of Riparian Right

According to this doctrine, the owners of lands abutting a river have an equal right to the use of the waters of this river, as long as they do not interfere with the rights of their co-riparians. Each riparian owner has a right to have the water flow past his or her land undiminished in quantity and unimpaired in quality. This doctrine has never been accepted as a basis for the solution of international water law disputes (Caponera 1994, 12). It is practiced among the Palestinians in both the West Bank and Gaza and aggravates the severe situation of groundwater resources.

The Prior Appropriation Doctrine

This doctrine asserts that water in its natural course is public property and is not susceptible to private ownership. The right to its use may be acquired by appropriation and application to beneficial use. The first appropriator establishes a prior right to the use of the water, always provided that this water is put to beneficial use (Caponera 1994).

The Theory of Absolute Territorial Sovereignty

According to this theory, also known as the "Harmon Doctrine," each state, because of the absolute sovereignty it exercises over its territory, may use waters of an international river within such territory as it pleases, without regard to the damage which may be caused to the other riparian states (Caponera 1994, 13; Kliot 1994, 5). There is general agreement nowadays to reject this doctrine as contradictory to customary international law. According to Dellapenna (1995), within the context of the Jordan-Yarmuk basin, this doctrine was espoused by Lebanon, the Jordanians before 1967, and Syria in the Jordan-Yarmuk basin, but not in the Euphrates and Orontes basins, in which it is a lower riparian (Dellapenna 1995, 279). Israel also adopted this doctrine when, in 1967, it became an upper riparian of the Jordan River.

The Doctrine of Absolute Territorial Integrity

Lower riparians often favor the principles of absolute territorial integrity, in which no state may utilize the water of international rivers in a way which might cause any detrimental effects on co-riparian territory. The Israelis, dependent as they originally were on water flowing down from Syria and Lebanon, initially adopted that doctrine. In 1967 they changed their position when their relative location on the river improved.

The Theory of Equitable Apportionment

According to this theory each co-basin state is entitled to a fair share of the waters of an international river or basin, if this entitlement is justified (Caponera 1994, 13). Improvement of this principle is expressed in the doctrine which is the most accepted today, the theory of equitable utilization (see below).

The Theory of Limited Territorial Sovereignty

This is a step which limits the sovereignty of states in international river basins. It recognizes the existence of a community of interests among riparian states that gives rise to a series of reciprocal rights and obligations (Caponera 1994, 14).

The Theory of Equitable Utilization

This is the commonest doctrine of customary international law in connection with international water resources. It is reflected in two similar sets of rules which were developed by international lawyers. The first set of rules are the Helsinki Rules, 1966, in which equitable utilization of international river basins is shaped by the following eleven points:

1. The geography of the basin, including the size of the drainage area in the territory of each basin state;

2. The hydrology of the basin;

3. The climate affecting the basin;

4. The past utilization of the waters of the basin;

5. The economic and social needs of each basin state;

6. The population in each state dependent on the waters of the basin;

7. The comparative costs of alternative means of satisfying the economic and social needs of each basin state;

8. The availability of other resources;

9. The avoidance of unnecessary waste in the utilization of waters of the basin;

10. The practicability of compensation to one or more of the co-basin states;

11. The degree to which the needs of a basin state may be satisfied without substantial injury to another basin state (United Nations 1970, 78–80).

Later on, the Helsinki Rules were strengthened by the Seoul Rules, which related to the prevention of pollution of international groundwater, the exchange of relevant information concerning aquifers, and integrated management of those resources (Kliot 1994, 8).

In the late 1980s the United Nations–affiliated International Law Commission began work on a similar set of rules. The I.L.C. Rules were submitted to the UN General Assembly in 1991 and are in an advanced stage of ratification by member states of the United Nations. The I.L.C. Rules concerning equitable utilization of shared water resources are similar to the Helsinki Rules and include these relevant factors:

1. Geographic, hydrographic, hydrological, climatic, biological, and other factors of a natural character;

2. The social and economic needs of the watercourse states concerned;

3. The effects of the use or uses of the watercourse;

4. Existing and potential uses of the watercourse;

5. Conservation, protection, development, and economy of use;

6. The availability of alternatives, of corresponding value, to a particular planned or existing use.

The I.L.C. Rules also stress the obligation not to cause appreciable harm (to other watercourse states), a general obligation to cooperate, a regular exchange of data and information, and giving priority to human consumption over any other uses of water in the international basins (Kliot 1994, Appendix B). Many of the above rules are reflected in the practices of states as documented in treaties and agreements among them (some of which are discussed below).

How is this body of customary international laws reflected in the agreements which Israel, the Palestinians, and Jordan have signed?

Israel and the Palestinians

The Declaration of Principles signed in Washington, D.C., on 13 September 1993, intending to form the basis of a peaceful settlement between Israel and the Palestinians, did not mention water, but Annex III to the declaration listed "equitable utilization of joint water resources" as a principle which would guide them in and beyond the interim period (Dellapenna 1995, 263). In the Water and Sewage Agreement (agreed version, 18 September 1995) Israel and the Palestinians reached an agreement for the interim period. The agreement does not make any specific reference to "equitable utilization" but adopts many provisions which could be clearly interpreted as representing this principle. For example:

3b. Preventing the deterioration of water quality in water resources;

3d. Adjusting the utilization of the resources according to variable climatological and hydrological conditions;

3e. Taking all necessary measures to prevent any harm to water resources. Israel also acknowledged the Palestinian domestic needs for water and agreed to immediate measures to increase water availability to the Palestinians (Article 7a).

The establishment of the Joint Water Committee—with responsibilities for coordinated management of water resources and systems, for exchange of information, for resolution of water and sewage disputes, for

arrangement of water supply from one side to the other—clearly demonstrates that both sides have adopted the norms of customary international law as reflected in the I.L.C. Rules and the practice of states.

The Peace Treaty between Israel and Jordan

The Israelis and Jordanians reached agreement on a Common Agenda
to serve as a framework for a peace treaty between them in November
of 1992. The Common Agenda states that securing the rightful water
shares of the two sides and searching for ways to alleviate water shortages are their guidelines, and the two will cooperate on water issues
(Dellapenna 1995, 265). The Treaty of Peace between Israel and Jordan,
Article 6, deals with water issues. There is mutual recognition of the
rightful allocations to both countries from the Jordan, the Yarmuk, and
the Wadi Araba-Arava groundwater. The parties have adopted a policy
of cooperation for developing new water resources and preserving the
existing resources, transfer of information, and mutual assistance in the
alleviation of water shortages. Annex II of the agreement allocates water
from the common water resources in a manner in which Jordan could
gain immediate relief from the water scarcity in its urban centers. There
are a wide variety of measures to prevent harm or damage to the common resources, arrangements for joint monitoring of common water resources, arrangements for data exchange, and the establishment of a
Joint Water Committee.

Again, many of these stipulations could be interpreted as falling into
the category of the "equitable utilization" doctrine of international law
concerning international water resources.

Features of Institutions for Joint Management of Water Resources

Another part of institutional frameworks for dealing with shared water
resources has to do with the institutions themselves. Here we differenti-
ate among three broad categories:

1. Agreements by riparian states stopping short of formal allocation or
   joint management of shared waters.

2. Agreements allocating water between states.

3. Agreements for joint (communal) management of internationally
   shared waters (Dellapenna 1995, 282–284).

Each of the three categories will be discussed, and its appropriateness to the shared Israeli-Arab water resources will be analyzed.

The first category embraces agreements in which there are provisions for an exchange of views without commitment to joint action, to exchange data, to monitor and talk, and to coordinate on specific issues (Eaton and Eaton 1994, 141). An example of an agreement in this category is the one regarding the Amur River basin and the Argun River. In it, Bangladesh and India called for joint research operations to determine the natural resources of the Amur and Argun Rivers, but the Joint Rivers Commission they established, a platform for an exchange of views, was unable to prevent the ensuing conflict between the two states. The Hydromet Project in the Upper Nile is also founded on collection, processing, and exchange of information. The Israeli-Palestinian Agreement and the Israeli-Jordanian Peace Treaty have provisions for exchange of information and views, which is just one of the functions assigned to the Joint Water Committees they create.

The second category of agreements is a more important one, as it either allocates water quotas from the common water resources or divides the water resources between the parties. Egypt and the Sudan, the downstream riparians of the Nile, decided to partition the water of the Nile between them in 1959. That treaty does not take into account the needs of the other riparians of the Nile—the equatorial states. India and Pakistan, the riparians of the Indus, after a long conflict decided to partition the river tributaries between them. The joint commission in this case supervises the full implementation of the treaty and intervenes in dispute resolution.

The settlement of the Indus dispute in the Indus Treaty of 1960 established the division of the river's water into two parts: the western tributaries were ceded to Pakistan (79 percent of the total volume of the water) and the eastern tributaries to India (Mehta 1986; Baxter 1967). The treaty established the Permanent Indus Commission. Each of the parties nominated a commissioner for the Permanent Indus Commission. Their function was "to establish and maintain cooperative arrangements for the implementation of the Treaty and to try to resolve by agreement any difference that might arise" (Sharma 1990, 197). As defined in the treaty, the commission's functions were: to study and report to the two governments on any problem relating to the development of the waters of the rivers; to make prompt efforts to settle any question arising therefrom; and to undertake, once in every five years, a general tour of inspection of the rivers to ascertain the facts concerning the various developments and works on the rivers. The commission was required to meet regularly at least once a year, in India and Pakistan alternately (Rahim 1990, 207). The treaty also provided for the regular exchange of data on discharge and extractions of water. It is interesting to note that though the history of conflict in the Indus is similar to that of hostility in the Middle East, the agreements between Israel and the Palestinians, and Israel and Jordan, are clearly more cooperative in their structure and contents and call for genuine cooperation in managing the common resources, while the Indus Treaty limits the cooperation to very specific issues.

The Israeli-Jordanian Peace Treaty, most particularly because of Annex II, which specifies water allocations, can be classified as a typical "allocation" agreement. There are also provisions for cooperation in building storage, protection of water resources, and other matters, but all fall short of the structure of a joint management agreement.

The Israeli-Palestinian Agreement also belongs to the category of allocation agreements, but the Joint Water Committee has wider authority which may qualify as "joint management"—for example, the establishment of supervision teams, making arrangements for supplying water from one side to the other, approval of licensing and drilling of new wells, and so on.

In a series of workshops which took place in 1994, 1995, and 1996, organized by Palestinian and Israeli water specialists and focused on joint management of shared aquifers, various aspects of joint management practices were discussed. Many of the participants took the view that a sequential institution-building process was desirable and that it need not lead to integrative structures. Also, institutions should be structured so as to allow confidence to be built over time, and monitoring could serve as a good starting point for building joint management programs. Pollution control, resource conservation, risk management, and training were the functions mentioned most often by participants as possible second steps toward comprehensive joint water management bodies. There was a feeling that the financial difficulties of the Palestinians would become a major impediment to the necessary infrastructure development. And, of course, the political acceptability of joint water management was questioned, a factor that the reality of 1998 points to as the most crucial one (Haddad and Feitelson 1995, 316–317).

The agreements in the third category, agreements for joint management of water resources, are considered the most appropriate for surface and groundwater resources, which are used for many functions and, often, for contradictory uses such as irrigation, power generation, and supplying water for industry. Multipurpose, basinwide institutions are the best frameworks for joint management of shared resources. There are some prominent examples of such frameworks. In Africa, the Senegal River Basin Development Authority (OMVS) is shared by four equatorial states. The Senegal River is used for irrigation, domestic needs, power generation, and navigation. The OMVS jointly owns and operates structures which it has built: two dams, a barrage port, and port installations (Godana 1985). The structure of the OMVS points, perhaps, to the reason it was able to perform tasks of genuine joint management. There were two basic conventions, those of 1972 and 1978, which established the current structure. The conventions established the joint ownership by member states of the two dams and the river seaports. The OMVS is responsible for promotion, coordination of studies, and construction works for the development of the Senegal. There are four components in the structure: (1) the Conference of Heads of State of Government, which determines policies; (2) the Council of Ministers, which is responsible for overall planning; (3) the High Commission, which is an executive body; and (4) the Permanent Water Commission, which defines the principles for the distribution of the Senegal water among states and uses. Perhaps such a structure, in which the highest level of leaders is involved, is the secret of this body's success.

But perhaps one of the best examples of joint management institutions is the International Boundary and Water Commission (IBWC), which is shared by the United States and Mexico. Its functions have steadily broadened, developing from a single purpose of boundary management to multifaceted resource management. The 1944 treaty concluded by the United States and Mexico concerning the Colorado and Rio Grande was basically an allocation agreement, but it determined priorities for utilization of the shared waters as follows: domestic and municipal use, agriculture, electricity generation, navigation, and fishing. Though the 1944 treaty did not give the IBWC powers in matters such as water quality and salinity, it was able to outline specific measures for the United States to undertake in order to reduce salinity in the water reaching Mexico, or problems of sanitation along the common boundary. The IBWC currently addresses issues of water quality, mainly by carrying out the engineering and scientific works, while either the United States or Mexico (depending on the benefits) covers the expenses. By adding minutes to the 1944 treaty the parties were able to deal successfully with issues such as groundwater allocation and competition among different uses. Although the United States and Mexico differ in their economic development, cultural background, water infrastructure, and process of decision-making, they are able to cooperate very successfully in the management of their shared water resources.

Why are agreements for joint management of international water resources preferred as the best framework for the task? Basically because any use of the basin- or groundwaters in one part has immediate repercussions or externalities on other parts. Therefore, the necessity to include all the co-basin states and all water uses under the auspices of such an institution. The success of these institutions depends solely on the type of powers delegated to them and the readiness of the partners to give up portions of their sovereignty over their water resources in order to manage them more equitably and efficiently.

Israelis, Jordanians, and Palestinians have manifested in the recent agreements, for the first time, a readiness to share common water resources more equitably and to pay full attention to water quality and sustainable management of these scarce and precious resources. The water provisions in both agreements reflect an enormous step ahead for the parties, who decided to end the state of war and their conflict over water resources and try to resolve the shortage by peaceful means. Basically, this means that the parties will be aided by the current body of international customary law concerning water resources and will adopt principles such as "equitable allocation and utilization of common water resources" or "obligation not to cause significant harm" in their newly evolved institutions. One can find clear traces of such principles in both the Israeli-Jordanian Peace Treaty and the Israeli-Palestinian Water and Sewage Agreement. However, the existing agreements exclude other riparians, and in order to prevent "significant harm" and, specifically, in order to enlarge the availability of water for all riparians, the consent and participation of all co-basin states are needed. Water importation, water storage, or projects such as the Med-Dead Sea Canal or Red-Dead Sea Canal necessitate the participation of all riparians. The agreement with the Palestinians adopted another principle of law by giving priority to the domestic use of the Palestinians, and that may improve the water supply for Palestinian urban centers, which suffer severe scarcity.

Perhaps it is unrealistic to ask that Israel, Jordan, and the Palestinians end their conflict by the adoption of one framework, endowed with broad powers, for joint management of their shared water resources. It is possible that when the permanent status of the Palestinian Authority is discussed, and after a period of building mutual confidence in the Israeli-Palestinian Joint Water Committee, the sides will be ready for a stronger and more authoritative institution. If and when Syria and Lebanon join the peace process, water issues certainly will be raised, and the framework for joint management of the Jordan-Yarmuk will necessarily have to expand. It helps to remember that, only a few years ago, the prospects for an Israeli-Palestinian accord or a peace treaty between Israel and Jordan seemed remote and unlikely.

CONCLUSIONS

Israel, Jordan, and the Palestinians share a history of hostility, war in rounds, and disputes over issues of "high politics," namely, legitimacy, borders, and raison d'être. When states are in dispute over issues of "high politics," they are disinclined to cooperate on issues of "low politics" such as the development and sharing of common water resources (Lowi 1993).

This chapter has presented the picture of the meager common water resources and history of competition over these resources, a few small "water wars," and unilateral, separate, and often contradictory development of the shared water resources. Many observers and commentators agreed that a war over these water resources was almost inevitable. Instead, the early 1990s brought with them a new era in the relations between Israel and its neighbors, and with this fresh era, new attitudes were adopted in connection with water resources. Though the new institutional frameworks, in the form of the Joint Water Committees, are in their initial steps, they represent the legal and political commitment of both Israel and its neighbors to a new legal regime to regulate water allocation and use.

REFERENCES

Abu-Sway, B., K. Al-Jamal et al. 1994. *Palestinian Water Resources*. Jerusalem: Water Resources Action Program.

Amery, H. 1993. "Shared Water Resources Management and Planning: Cooperative Water Management in the Middle East." In *Proceedings of the International Symposium on Water Resources in the Middle East: Policy and Institutional Aspects*, pp. 59–68. Urbana: University of Illinois at Urbana-Champaign, International Water Resources Association, 24–27 October.

Bakour, Y., and J. Kolars. 1993. "The Arab Mashrek: Hydrologic History, Problems and Perspectives." In P. Rogers and P. Lydon (eds.), *Water in the Arab World*, pp. 121–146. Cambridge, Mass.: Harvard University Press.

Baxter, R. R. 1967. "The Indus Basin." In A. Garretsor et al. (eds.), *The Law of International Drainage Basins,* pp. 443–489. New York: Oceanic Books.

Beschorner, N. 1992. *Water and Instability in the Middle East.* London: International Institute for Strategic Studies.

Caponera, D. 1994. "The Legal-Institutional Issues Involved in the Solution of Water Conflicts in the Middle East." In J. Isaac and H. Shuval (eds.), *Water and Peace in the Middle East,* pp. 163–180. Amsterdam and London: Elsevier.

Dellapenna, J. 1995. "Designing the Legal Structures of Water Management Needed to Implement the Israeli-Palestinian Declaration of Principles." In M. Haddad and E. Feitelson (eds.), *Joint Management of Shared Aquifers: The Second Workshop,* pp. 261–310. Jerusalem: Palestine Consultancy Group and Harry S. Truman Research Institute.

Eaton, D., and J. Eaton. 1994. "Joint Management of Aquifers between the Jordan River Basin and the Mediterranean Sea by Israelis and Palestinians: An International Perspective." In M. Haddad and E. Feitelson (eds.), *Joint Management of Shared Aquifers: The Second Workshop,* pp. 131–152. Jerusalem: Palestine Consultancy Group and Harry S. Truman Research Institute.

Godana, B. A. 1985. *Africa's Shared Water Resources.* Boulder, Colo.: Lynne Rienner.

Gruen, G. 1993. "Recent Negotiations over the Waters of the Euphrates and Tigris." In *Proceedings of the International Symposium on Water Resources in the Middle East: Policy and Institutional Aspects,* pp. 100–107. Urbana: University of Illinois at Urbana-Champaign, International Water Resources Association, 24–27 October.

Haddad, Marwan, and Eran Feitelson. 1995. *Joint Management of Shared Aquifers: The Second Workshop.* Jerusalem: Palestine Consultancy Group and Harry S. Truman Research Institute.

Hof, F. C. 1995. "The Yarmuk and Jordan Rivers in the Israel-Jordan Peace Treaty." *Middle East Policy* 3 (3): 47–56.

Inbar, M., and J. Maos. 1984. "Water Resources Planning and Development in the Northern Jordan Valley." *Water International* 9 (1): 18–25.

Israel, Government of. 1994. *Treaty of Peace between the State of Israel and the Hashemite Kingdom of Jordan 26 October 1994.* Jerusalem: Ministry of Foreign Affairs.

————. 1995. *Interim Agreement with the Palestinian Authority—Article 40. Water and Sewage.* Jerusalem: Ministry of Foreign Affairs.

Jordanian Times. 1994. Editorial. 20–21 October.

Kahan, D. 1987. *Agriculture and Water Resources in the West Bank and Gaza 1967–1987.* Jerusalem: West Bank Data Base Project.

Kahana, Y. 1994. "The Turonian-Cenomanian Aquifer: The Need for a Joint Monitoring and Management Program." In M. Haddad and E. Feitelson (eds.), *Joint Management of Shared Aquifers: The Second Workshop,* pp. 1–22. Jerusalem: Palestine Consultancy Group and Harry S. Truman Research Institute.

Kally, E. 1986. *A Middle East Water Plan under Peace.* Tel Aviv: Tel Aviv University, Armand Hammer Fund for Economic Cooperation in the Middle East.

Kliot, N. 1994. *Water Resources and Conflict in the Middle East.* London: Routledge.

Lindholm, H. 1992. "Water and the Arab-Israeli Conflict: An Imperative for Regional Cooperation?" In Leif Ohlsson (ed.), *Regional Case Studies of Water Conflicts*, pp. 46–83. Göteborg: University of Göteborg Pedigru Papers.

Lonergan, S., and D. Brooks. 1993. *The Economic, Ecological and Geopolitical Dimensions of Water in Israel*. Victoria, B.C.: University of Victoria Centre for Sustainable Regional Development.

————. 1994. *Watershed: The Role of Fresh Water in the Israeli-Palestinian Conflict*. Ottawa: International Development Center.

Lowi, M. 1993. *Water and Power: The Politics of Scarce Resources in the Jordan River Basin*. Cambridge, UK: Cambridge University Center.

Mehta, J. S. 1986. "The Indus Water Treaty." In E. Vlachos, A. Webb, and I. Murphy (eds.), *The Management of International River Basin Conflicts*, pp. 1–24. Proceedings of a Workshop held at the Institute for Applied Systems Analysis, Laxenberg, Austria, 22–25 September. Laxenberg, Austria: Institute for Applied Systems Analysis.

Murakami, M., U. El-Hanbali, and A. Wolf. 1995. "Technological Alternative Strategies in Interstate Development of the Jordan Rift Valley beyond the Peace." *Water International* 20 (4): 188–196.

Murakami, M., and K. Musaike. 1995. "The Jordan River and the Litani." In Asit K. Biswas (ed.), *International Water of the Middle East*, pp. 117–155. Delhi, India: Oxford University Press.

Naff, T. 1992. "Water Scarcity, Resource Management and Conflict." In *Jordan's Water Resources and Their Future Potential*, pp. 107–112. Proceedings of the Symposium organized by Friedrich Ebart Stiftung and the Water Research and Study Centre, University of Jordan, Amman, Jordan, 28–29 October 1991.

Nassereddin, T. 1994. "Institutional Aspects of Joint Management of Aquifers by Israel and the Occupied Territories." In M. Haddad and E. Feitelson (eds.), *Joint Management of Shared Aquifers: The Second Workshop*, pp. 60–67. Jerusalem: Palestine Consultancy Group and Harry S. Truman Research Institute.

Rahim, A. 1990. "Indus Basin Development." In *United Nations Department of Technical Cooperation—Natural Water Resources Series No. 20, River and Lake Basin Development*, pp. 203–210. Proceedings of the UN Meeting, Addis Ababa, Ethiopia, 10–15 October 1988. Addis Ababa: United Nations.

Salameh, E. 1992. "The Jordan River System." In *Jordan's Water Resources and Their Future Potential*, pp. 99–106. Proceedings of the Symposium organized by Friedrich Ebart Stiftung and the Water Research and Study Centre, University of Jordan, Amman, Jordan, 28–29 October 1991. Amman, Jordan: University of Jordan.

Sharma, K. S. 1990. "India's Experience in Developing the Indus River Basin Program." In *United Nations Department of Technical Cooperation—Natural Water Resources Series No. 20, River and Lake Basin Development*, pp. 194–202. Proceedings of the UN Meeting, Addis Ababa, Ethiopia, 10–15 October 1988. Addis Ababa: United Nations.

Shuval, Hillel. 1992. "Approaches to Resolving the Water Conflicts between Israel and Her Neighbors: A Regional Water for Peace Plan." *Water International* 17 (3): 133–142.

Stevens, G. 1965. *The Jordan River Partition.* Stanford, Calif.: Hoover Institution.

Taubenblatt, S. A. 1988. "Jordan River Basin Water: A Challenge in the 1990's." In J. Starr and D. Stoll (eds.), *The Politics of Scarcity: Water in the Middle East,* pp. 41–52. Boulder, Colo.: Westview.

United Nations. 1970. *Integrated River Basin Development.* New York: United Nations Department of Economic and Social Affairs.

Al-Weshah, Radwan al-Mubarak. 1992. "Jordan's Water Resources: A Technical Perspective." *Water International* 17 (3): 124–132.

Wolf, A. T. 1995a. *Hydropolitics along the Jordan River.* Tokyo and New York: United Nations University Press.

———. 1995b. "International Water Dispute Resolution: The Middle East Multilateral Working Group on Water Resources." *Water International* 20 (3): 141–150.

———, and M. Murakami. 1995. "Techno-Political Decision Making for Water Resources Development: The Jordan River Watershed." *Water Resources Development* 11 (2): 147–162.

World Bank. 1994. *Integrated Development of the Jordan River Valley.* Washington, D.C.: World Bank.

9. Political Controls of River Waters and Abstractions between Various States within the Middle East
LAWS AND OPERATIONS, WITH SPECIAL REFERENCE TO THE JORDAN BASIN

Gwyn Rowley

INTRODUCTION

Over a third of the world's population lives in river basins divided between two or more countries. India and Bangladesh haggle over the Ganges, Mexico and the United States over the Colorado, Slovakia and Hungary over the Danube, and Thailand and Vietnam over the Mekong. Nowhere, however, are water disputes shaping political landscapes and economic futures more dramatically than within the Middle East. Indeed it has been asserted that "Over the next decade, water issues in the three major river basins will foster an unprecedented degree of co-operation or a combustible level of conflict" (Postel 1992, 46).

In these areas, increasing populations, expanding agriculture, and mounting industrialization, matched by correspondingly higher living standards, demand more fresh water, while drought and pollution, war and mismanagement, limit its availability. However, it must be recognized that "In this patchwork of ethnic and religious rivalries, water seldom stands alone as an issue; it is entangled in the politics that keep people from trusting and seeking help from one another" (Vesiland 1993, 140).

Here within the Middle East water is more than "a resource"; it confers power and wealth for those who control it, underdevelopment and poverty for those who do not. Such highly competitive conditions have led many to predict that the next war in this sensitive and volatile region could well be over water (Pugh 1990; Irani 1991; Amery 1993). Yet it has been suggested that "The possibility for mutual gain from co-operation exists for all nations dependent on river basins" (Smith 1966, 125).

However, at present the chances for voluntary cooperation on certain major issues concerning water resources within the Middle East seem highly unlikely because of historical animosities between neighboring states and their continuing egotistical self-interests.

As yet, international law appears to offer minimal assistance in resolving such conflicts, despite a number of attempts by various international agencies. This present study focuses on certain of these attempts within the Middle East, specifically those in the Jordan basin. The case of the Colorado basin within the United States will be used as a comparison. I also consider the likelihood of the "success" of more general solutions to the problem. The question as to why no definite legal measures have yet been developed is also considered. The study offers outline suggestions on possible steps for the future, to try to resolve the potential crisis, and certain of the tangible gains that could result for all parties from a greater and more flexible cooperation in a stabilized Jordan basin. Yet this also has to be considered against the background of the continuing expansionist Israeli settlement policies throughout most of the Occupied Territories (Rowley 1990a; 1999).

Certain of the other contributors to this volume may possibly present an overtly optimistic view on the realization of such significant compromises amongst Israel, the Palestinians, and its other Arab neighbors on water issues within the Jordan basin. My personal view is somewhat more reserved and differentiates amongst the several parts of the basin and between the shorter and longer terms. The base question here, in essence, relates to how further compromises on water issues would benefit Israel, being ever mindful that Israel is the dominant military power within the Jordan basin and is consistently supported by the United States.

WATER AS A STRATEGIC RESOURCE

Certain resources are defined as "strategic" because they are both vital and scarce. These include petroleum, and certain metals, such as iron ore and cobalt, but probably most important, yet generally less appreciated, is water. These resources have a potential to create resource geopolitics. Water is most scarce in the arid and semiarid regions of the world, and Elliot (1991) and Postel (1992) isolate the Middle East as that region which is most vulnerable to such shortages. Annual rainfall, apart from the far north about Turkey and eastward to the Caspian Sea, and in the deep south about the headwaters of the Nile, is low, generally ranging from 100 to 400 mm, depending on the relief. It is also variable (see Thornthwaite, Mather, and Carter 1958). The scarcity has been exacerbated by the high rates of population increase, with population set to double over the next twenty years, and the resulting need for economic development in both agriculture and the secondary and tertiary sectors of the economy. The desire for self-sufficiency in food has shown itself in the high levels of construction of water storage facilities and exploration for ever more resources. It has been suggested that "By the year 2000, only three countries in the Middle East—Iran, Turkey and Syria—will have a per capita consumption of water safely above the accepted minimum" (Anderson 1991a, 32).

However, the main reasons why the threat of "water wars" (Elliot 1991) is so acute is because the major rivers flow through more than one country and most of the countries within the region appear to detest each other.

Examples of conflicts in the three major basins within the Middle East, the Jordan, the Nile, and the Tigris-Euphrates, include that due to the Syrian Thawra Dam in 1974, which cut the flow of the Euphrates into Iraq by some 25 percent. Iraqi troops were sent to the border, and open conflict and bloodshed seemed imminent until Syria backed down. The most recent potential conflicts surrounding the Tigris-Euphrates basin derive from the Southeast Anatolia Project (GAP) in Turkey, which, following the completion of its twenty-two dams, could cut Syria's share of the Euphrates waters by some 40 percent and Iraq's share by 80 percent.

Within the Nile basin, where a fragile control is divided amongst the nine Nilotic nations, the most recent threats to peace come from Ethiopia, which has proposed to dam the origin of the Blue Nile. As this provides some 86 percent of the flow of the Nile, it therefore "holds the key to potential Nile hydropolitics" (Anderson 1991b, 13). Past conflicts also include that over the Aswan High Dam, where military confrontation occurred in 1958 between Egypt and Sudan (Allan 1992; Rowley 1993). The current Toshka basin project to develop the New Valley westward of Lake Nasser will create further demands upon Nile waters. However, it is in the Jordan basin, by far the smallest of the three major river basins within the Middle East, where the potential for geopolitical activity and increasingly wider conflict is now most apparent.

CONFLICT IN THE JORDAN BASIN

The Jordan basin is shared by four states: Jordan with 54 percent of its territorial area, Israel with 10.5 percent, Syria with 29.5 percent, and Lebanon with the remaining 6 percent. Jordan itself controls no major rivers, and whereas the Jordan River forms part of its border with Israel, the river's headwaters rise in the Lebanon, Israel, and Syria. Therefore,

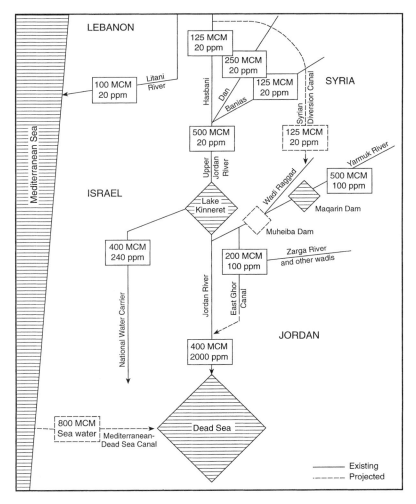

FIGURE 9.1. A schematic diagram of the Jordan and lower Litani catchment areas, indicating the approximate annual flows in million m³(mcm) and salinities in parts per million (ppm). Redrawn from Rowley 1984.

the Hashemite Kingdom depends on its main tributary, the Yarmuk, but this is also crucial to both Syria and Israel (see Figures 9.1 and 9.2).

As Elliot has noted: "In an ideal world the countries in the region would work out a way of sharing their water. In reality the tortured politics of the Middle East have made a bad situation worse" (1991, 30).

Major conflicts came to the fore with the creation in 1948 of Israel, which then, almost immediately, set in motion plans to divert the upper

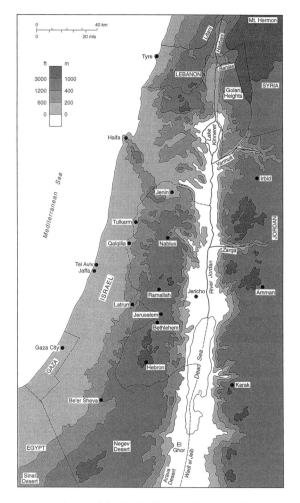

FIGURE 9.2. A general map of the Jordan Basin and surrounding areas.

Jordan waters into the Israeli National Water Carrier (NWC) scheme (see Smith 1966).

The NWC, a 270-mile-long (432-km-long) canal and pipeline to transport upper Jordan water in a general southwesterly and then southerly direction through Israel toward the Negev, resulted in the movement of saline water into the lower Jordan. Jordan and Syria were understandably outraged, calling the pipeline a breach of international law. Israel, however, maintained that it had the right to do what it wanted with its own water: "In effect, Israel has recreated the River Jordan as a pipeline in its own territory" (Pearce 1991, 37).

"Full utilization" of the Jordan water has been said to be hydrologically possible (Smith 1931; Smith 1966, 111), and in 1953 an attempt was made by the United Nations and the United States to get Israel and the Arab states to agree to a unified river basin development scheme, which would take into account existing boundaries and rival demands.

This plan formed the starting point for the negotiations which Eric Johnston, President Dwight Eisenhower's special representative, held with Israel and the Arab states between 1953 and 1955. The final water apportionment figures put forward, however, were rejected by both sides, Israeli and Arab, "even though technically, based on strict geographical and hydrological factors, it was the best possible outcome" (Smith 1966, 123).

The outline details of the envisaged settlement are presented as Table 9.1. However, the Arab side complained that Israel would receive 33 percent of Jordan water despite having only 23 percent of the river. The Israelis counterclaimed that there was in fact 2,345 mcm per annum of water flow in the basin (as opposed to the U.S. estimate, which stated there was only a total of 1,452 mcm) and demanded rights to 1,240 mcm (see Kahhaleh 1981).

The Johnston Plan has nonetheless been used as a guideline for the use of the Jordan waters ever since; cooperation over water is therefore certainly possible.[1]

However, with no official agreement having been reached, both Israel

TABLE 9.1. Original Proposal, Arab Proposal, and Final Guidelines for Use of Jordan Basin Waters Set in 1953–1955 (million m^3)

	Original Proposal	Arab League Proposal	Final Guidelines
Jordan	774	977	720
Israel	394	285	565
Syria	43	132	132
Lebanon	0	35	35
Totals	1,213	1,429	1,452

SOURCE: Kahhaleh (1981).

and Jordan continued with their own water diversion schemes, and "the cornerstone of Israel's strategy was then to take by force what it could not get through negotiation" (Jaber 1989, 66).

Today, although the Jordan basin is fraught with water-related tensions, five crucial hydropolitical flashpoints may be isolated. These in turn point to the crucial relationship between power and water utilization in general.

Firstly, political tension over the Yarmuk has heightened since the Six-Day War of June 1967, when Israel captured and occupied the Golan Heights. This has made Israel a co-riparian of the Yarmuk, which gives it a direct interest in the Jordan, and also Syria's "Unity Dam" project (Anderson 1991b). Secondly, Israel's occupation of the Golan Heights gives it control over the Banias, another of the upper Jordan's tributaries. This control ensures that no non-Israeli scheme for the diversion of the Jordan headwaters can be implemented.

Thirdly, the Israeli invasion of the Lebanon in June 1982 resulted in the creation of an Israeli occupied zone in the southern Lebanon, which, while supposedly for security reasons, conveniently includes the southern section of the Litani River, the major river of the Lebanon (Hudson 1970). It also includes the Hasbani (Nahal Senir in Israel), which rises in a series of large springs on the western slopes of Mount Hermon (Figure 9.3). The Hasbani also receives additional water from the Ghajar (Wazzani) Springs, which lie some 3 kilometers north of the Israeli border (see Smith 1966). Control of Litani and related upper Jordan waters had been a long-standing Israeli intention, not only for their quantity and reliability, but also for their quality. Their low salinity is particularly attractive (see Figure 9.3).

That part of the Lebanon southward from the Litani River, shown in Figure 9.3, is presently a part of the Israeli Security Zone, while the territory defined in the figure as "Syria" is now part of the Israeli-occupied ("annexed") Golan Heights.[2]

A continuing Israeli presence within the southern Lebanon and the western Golan is here considered as crucial to present Israeli water schemes within both the upper Jordan Valley and throughout Israel (see Rowley 1984, 144–146). It will be most interesting to follow longer-term Israeli plans for the movement of Litani waters into the Jordan basin.

The fourth flashpoint relates to the Israeli occupation of the West Bank and its major aquifer (Rowley 1990b). According to some reports Israel is using 90 percent of all water abstracted from the West Bank, which in turn is contributing to a decline in the water table and a desic-

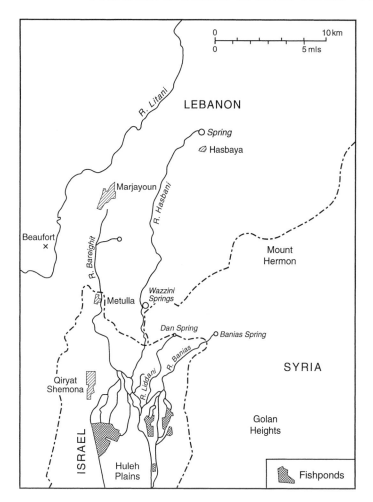

FIGURE 9.3. The upper Jordan Valley. Source: Smith 1966; Israel Defense Forces Spokesman 1981a; 1981b.

cation of many local Arab springs and wells (Rowley 1990b). To place this into the broader context, about a quarter of Israel's annual water potential originates from the West Bank (Benvenisti, Abu-Zayed, and Rubinstein 1986, 223–225).

The fifth and final matter relates to the fourth. Whereas the "peace process" figures prominently in the reporting of the Israeli-Palestinian issue, matters on the ground are somewhat more complex. Rowley (1999) considers the Israeli settlement program proceeding throughout the Occupied Territories, including East Jerusalem. Such colonialist expansions include the Seven Stars Plan, to extend Israeli state-space in the Triangle—that area between Tulkarm in the north and Latrun in the south—to push Israeli space farther eastward into the Occupied Territories. Settlement also proceeds under the Judean Hills Initiative in the Gush Etzion settlement block, northward from Hebron (Al-Khalil in Arabic). Further Israeli settlement and consolidation continue throughout the remainder of the West Bank, with a continuing Israeli civil and military presence within Gaza, and the deepening and creeping Judaization of East Jerusalem. Such a program points to further confirmation of Israel's overall control of land, routeways, and water resources throughout most of the Occupied Territories (Rowley 1990c; 1992).

Yet such proceeding developments in turn serve seriously to threaten and indeed undermine the Oslo Accords of 1994, brokered by the United States, between the Israeli government and the Palestinian Liberation Organization (see Newman 1994). It is against this backcloth of the continuing Israeli-Arab conflict, as briefly sketched out above, that any deeper evaluation of the possible scenarios on Jordan basin water issues must be set. Here we must be mindful that some consider that:

> Since the creation of the state of Israel in 1948 the problem of an equitable division of the Jordan waters has been both a symptom of the relations between Israel and her Arab neighbours, and one of the most important and intractable problems preventing an Arab-Israeli settlement. (Smith 1966, 111)

Indeed, the issue of water in the Jordan basin seems to have evolved into a race to see who can get the most of this strategic resource. This race is currently being won by Israel, the most politically and militarily powerful nation within the Jordan basin. It therefore appears to be a case of the survival of the fittest, and unless something can be done to halt Israel's lead, sooner or later the combined desperation of the other riparians could well lead to war. The question remains as to what can be done. If agreement fails to work is there any conceivable recourse to a legal process?

OBSTACLES TO FORMING AN INTERNATIONAL WATER LAW

It has been pointedly asserted that "In the present state of affairs, in its fragmentary and undeveloped stage, international water resource law can, unfortunately, offer little concrete guidance to the international drainage basin states" (Chauhan 1981, 52).

Teclaff (1991) believes that the root of the problem stems from the fact that, from very early in history, states asserted power over all waters within their borders and guarded that power jealously. This, then, delayed the acceptance of any notion regarding the legal unity of the international river basin. Since the availability of flowing water was important for the well-being of the irrigated community, and the use of water was under tight state control, additional water for the expansion of irrigation could be obtained either by agreement or conquest. However, Teclaff suggests that "It may plausibly be argued that such lack of voluntary co-operation in water management led to attempts to conquer entire basins in arid areas and contributed to establishing empires based on these major river basins" (1991, 60).

Later the concept of customary law, which occurs where and when "a clear and customary habit of doing certain actions has grown up under the aegis of the conviction that these actions are obligatory and right" (Oppenheim 1955, 171), was transferred from local groups of riparian owners, which had a collective interest in how their waters were to be used, to states. However, although this introduced to international law the concept of a community of riparians, it failed to provide specific and clear, unequivocal guidelines from which rules for the use of international river basins could evolve. Therefore, laws regarding international river basins are now enacted according to one of four alternative theories, or *principles,* which govern the rights and obligations of co-riparian nations (Berber 1959, 13).

The first of these is the principle of absolute territorial sovereignty. Here a state may dispose freely of waters that actually flow through its territory, but it has no right to demand the continued free flow from other countries. This operates practically only in favor of the uppermost riparian within a river basin. Secondly, there is the principle of absolute territorial integrity, by virtue of which a state has the right to demand the continuation of the natural flow of water coming from other countries, but may not, for its part, restrict the natural flow of water flowing through its territory and other countries. This principle operates practically only in favor of the lowest riparian.

These first two principles, according to Berber, do not provide a solution to international water disputes, "[as] they are grounded in an individualistic and anarchical conception of international law in which personal and egotistical interests are raised to the level of guiding principles and no solution is offered to the conflicting interests of the upper and lower riparians" (1959, 13–14).

Thirdly, there is the principle of community in the waters, by virtue of which rights are either vested in the collective body of riparians or are divided proportionally; or any other kind of absolute restriction on the free usage of the waters by the riparians is created in such a way that no one state can dispose of the waters without the positive cooperation of the others.

The community in the waters principle thus relies on the natural law principles of "good faith and good neighbor relations" and is well known in municipal water rights. In essence, the underlying sentiment is that "The gifts of nature are for the benefit of mankind and no aggregation of men can assert and exercise such right and ownership of them as will deprive others of having equal rights and means of enjoyment" (Farnham 1904, 56). However, Berber (1959) questions whether the international community is developed and mature enough for this to happen.

The fourth principle, the principle of restricted territorial sovereignty and restricted territorial integrity, is an elaboration and modification of the third, more suitable to a less advanced level of international integration. It relates to a restriction of the free usage of the water which, it is true, does not extend as far as the principle of a community in the waters but which, in differing degrees, restricts the principle of absolute territorial sovereignty just as much as the principle of absolute territorial integrity. This more lenient principle seems to be what most teaching on the subject conforms to today. However, because of its breadth, it is likely that its effect will be less distinct than that of the other principles and will probably contain obscurities and inaccuracies.

The general acceptance of the first principle is emphasized by the upstream states in the Middle East. This is the crux of the problem in the Tigris-Euphrates basin. Turkey is the upstream state, and it firmly believes that downstream countries have no right to interfere with Turkish internal water policy (Rowley 1993, 194–195). As Turkish President Suleyman Demiral has stated:

> Why should they [Syria and Iraq] have rights to the waters of Turkey? Do we have the right to the oil of these downstream countries? The upstream people have the absolute right to use this water. The Turkish waters are not international waters. (quoted in Irani 1991, 25)

The further aspect, of course, is the military might of Turkey, Egypt, and Israel within their respective river basins, the Tigris-Euphrates, the Nile, and the Jordan, providing a particularly volatile ingredient for water issues within the entire region that could serve further to undermine the already fragile agreements within these three basins (see Irani 1991;

Rowley 1993). Thus the very heart of the problem seems to be that water is not being treated as a universal resource, but as a national one which must be defended at all costs.

The concept of nationalism is also considered by Todd (1992) in a study of Texas groundwater law. Todd discusses the difficulties of efficiently and equitably defining, allocating, and protecting rights to a common fluid without guidance from publicly agreed and enforced rules. He specifically asks why more efficient guidelines or policies have not been adopted and comes up with a few suggestions of nonlegal factors which reduce the likelihood of agreement. These include ignorance; costs, both technical and legal; the limitations of self-help, for example, the difficulties of enforcing voluntary agreements; and the fact that the benefits of cooperation are difficult to show or are slow to appear. Finally, Todd considers the concept of the "Tragedy of the Commons." Ideally, when a nation uses water from a river it should weigh the opportunity costs of not having that water available in the future. This should act as a strong check on present use. However, where a country shares a water supply with another country, it can no longer be sure that unused water will remain for its use in the future, as the other country may use it. Todd concludes that "There is little reason therefore to forestall today's use, and the race begins" (1992, 250).

This has already been illustrated to be the case in the Middle East and as the factor which is currently causing most of the problems.

In a particularly engrossing study Chauhan (1981) considers the variations in conditions and circumstances within various international drainage basins in different parts of the world that make it difficult to lay down rigid and universal rules. He concludes that "Rules should take into consideration the peculiarities and various shades of geographical, legal, technical, sociological, economic and other conditions" (Chauhan 1981, 174). Is it possible for these difficulties to be overcome?

THE RIVER BASIN CONCEPT AND THE HELSINKI RULES

Perception of the interdependence and interrelationship of the water resources within a river basin is considered by a number of international organizations to provide the basis of "an answer." In fact, the underlying concept of the unity of a river was taken up in the closing decades of the nineteenth century by planners of multipurpose water developments, and was later extended to embrace entire river basins. In less than half a century jurists were able to arrive at a body of rules for the community of interests of states in a river basin. In 1911 the International Law Institute (ILI), in its Madrid Declaration, stated that "Riparian states with a common stream are in a position of permanent physical dependence on each other." The Institute drew up two essential rules resulting from such an interdependence which states should observe (see Macrory 1985). Firstly, when a stream forms a frontier of two states, "neither State may, on its own territory, utilize or allow the utilization of the water in such a way as to seriously interfere with its utilization by the other State or by individuals, corporations, etc. . . . thereof" (quoted in Teclaff 1991, 67). The second ILI rule stipulated that the essential character of the stream, when it reaches a downstream territory, should not have been seriously modified.

Some forty years later the International Law Association (ILA) put its stamp of approval on the notion of the integrated river basin as the proper unit for the cooperation of states in developing water resources. The Salzburg Declaration of 1961 then found that the rights of states to use waters flowing across their borders were defined by the principles of equity, but such principles were not specified. Thus doubts still persist as to whether there are any specific legal rules applicable to the waters of international river basins.

However there was no room for doubt in the ILA's Helsinki Rules, a crowning achievement of the ILA after many years of labor (Teclaff 1991, 69). Indeed, the Helsinki Rules of 1967 may be said to be the first attempt by an international organization to prepare a complete and all-embracing codification of the law of international watercourses.

The keystone of the Helsinki Rules is the concept of "the international drainage basin," which represented a significant step forward. McCaffrey declared that "For the first time it confirmed, in terms of juridical area, what naturalists, engineers and economists had previously accepted: that the basin as a physical whole and not only the river or the waters, must be the object of legal regulation" (1991a, 143). Several years later the International Law Commission (ILC) discovered that the use of the term "drainage basin" was, not surprisingly, unpopular with upstream riparians. Such states generally favored the use of the term "international river," a definition which excludes not only tributaries but also groundwater.

In an endeavor to overcome this problem the ILC introduced the phrase "international watercourse system," which constitutes a "shared natural resource." Reservations were again expressed concerning the terms "system" and "shared," because they were seen as similar to the drainage basin concept, and instead an article was proposed providing that a state is "entitled to a reasonable and equitable share of the uses of the waters" (McCaffrey 1991a, 157).

However, not everybody shares the view of Teclaff in his praise of the ILC. Chauhan (1981) had earlier suggested that, although rules have great utility, they cannot be claimed to provide a final and complete answer to all legal problems pertaining to uses of waters in international basins. Chauhan specifically pointed to the use of the phrase "equitable utilization," emphasizing that "This can only be seen as help material rather than a concrete answer in determining the exact respective share of a contested basin state" (1981, 196).

Postel (1992) also considered the principles proposed by the ILC: for example, the need to inform and consult with water-sharing neighbors, to avoid causing substantial harm, and the proposal of equitable utilization. She concluded that each of these aspects is open to different interpretations: "The factors to be considered in determining reasonable and equitable use are so many and so broad as to offer little practical guidance" (Postel 1992, 120).

Finally, Tarasofsky (1993) summarizes the results of a conference on water in the Middle East held in 1992. Here it was stated that the standards set by the ILC were "too elastic," and that this, and the importance placed on the role of negotiations, favored the interests of stronger states. He is particularly outspoken in pronouncing that "This result is due to political compromise by the ILC in the face of opposition by some upper riparian states" (Tarasofsky 1993, 71).

In addition, the 1992 conference concluded that it also felt that the ILC had left out important issues from its remit, such as issues related to large-scale diversions, roles for international organizations in promoting cooperation, and the problems of groundwater.

As McCaffrey (1991a) points out, the twenty-seven articles adopted by the ILC between 1987 and 1990 did include the term "system," but enclosed in brackets, to indicate that no final decision had been made. Also the expression "shared natural resource" was not included, and, in fact, the verb "share" did not appear anywhere. McCaffrey is equally outspoken when he remarks: "Thus it would appear that the ILC, largely in response to the views expressed by governments, has backed away from the express recognition of the interdependence of the various components of hydrographic systems" (1991a, 160).

It is concluded here that although the ILA and ILC did move one step further toward some legal agreement, it was not enough, and unless the states concerned begin to cooperate, and unless these organizations stop bowing to the governments of the politically stronger riparians, it is unlikely that any precise rules of law are ever going to be generally accepted. As the likelihood of cooperation between states that have an historical animosity toward each other is very unlikely, perhaps the creation of general rules is not the best way forward in the settlement of such disputes.

At this point some direct attention needs to be given to the recent *Draft Articles on the Law of the Non-Navigational Uses of International Watercourses* (ILC 1994). It is stressed that the entire tenor of this document is permissive rather than mandatory, and open to so many interpretations and value judgments that they render it quite meaningless both as a potential legal document or as a draft to act as a basis for further elaboration.

The title *The Law of the Non-Navigational Uses of International Watercourses* is therefore most inappropriate and indeed unacceptable.

My *Oxford English Dictionary* defines "law" in part as a

Body of enacted or customary rules recognized by a community as binding.

While my Webster's dictionary refers to

A rule prescribed by authority; a statute, a precept . . . legal procedures, litigation.

This draft report cannot be said to accord to such definitions of law.

Let us briefly consider the looseness and arbitrary nature of this document. Within the draft report there is a recurring use of the terms "watercourse" and "watercourse state"; the term "river basin" is not used. Thus the earlier criticisms of ILC drafting by McCaffrey (1991a), Postel (1992), and Tarasofsky (1993) would still appear to stand.

A focus on four articles out of the thirty-two within this draft should suffice. These are reprinted here in part:

Watercourse agreements
1. Watercourse States may enter into one or more agreements. . . . (Article 3, in part)
Management
1. Watercourse States shall, at the request of any of them, enter into consultations concerning the management of an international water-

course, which may include the establishment of a joint management mechanism. (Article 25, in part)

Indirect procedures

1. In cases where there are serious obstacles to direct contact between watercourse States, the States concerned shall fulfill their obligations of co-operation provided for in the present articles, including exchange of data and information, notification, communication, consultations and negotiations, through any indirect procedures accepted by them. (Article 31)

Data and information vital to national defense and security

Nothing in the present articles obliges a watercourse State to provide data or information vital to its national defense or security. (Article 31, in part)

Thus, in brief, Article 3 suggests that watercourse states *may* enter into one or more agreements; alternatively, they may not! Article 25 uses the more definite word "shall" relating to consultation, but again only the permissive "may" when referring to the establishment of a joint management mechanism. Likewise, neither Article 31, nor any of the other articles, either singlularly or collectively, appear to refer to any set legal procedure for possible noncompliance with the aforementioned articles. Finally, Article 31 provides the possibility for a legal loophole of momentous proportions, whereby any noncompliance could be argued under the cachet "vital to national defense or security" (see also McCaffrey 1991b).

THE SUITABILITY OF TREATIES AS A MODE OF SETTLEMENT

In spite of Chauhan's belief that, in the present circumstances, it is extremely difficult to lay down certain concrete, rigid, hard and fast rules for the settlement of international water law disputes, he considers that "It is incumbent to evolve certain norms in this field which could serve as concrete guidelines if not exact rules" (1981, 55).

Indeed, the view that legislation in international law implies treaty-making is an opinion generally favored by many. International legislation by treaty has been extremely successful, with several hundred international water treaties having been concluded over the last few decades, and these appear to have been such "good law" that it has been necessary only in very few cases to resort to international adjudication.

One example of the success of such a treaty is the Mexico-USA

Colorado Treaty of 1944. The Colorado River flows some 1,300 miles (2,080 km) from its headwaters in Wyoming to its mouth in the Gulf of Mexico in the Republic of Mexico. The river had been the cause of many bitter and protracted struggles because control of it meant potential wealth and prosperity; without water from the river a state might be condemned to become a desert, with the attendant destitution. Conflicts within the basin have included those between upper and lower riparians, amongst states of the upper basin (such as Colorado and Utah), amongst states of the lower basin (such as Arizona and California), and between the United States and the Republic of Mexico.

In 1924 the U.S. Congress passed an act authorizing the President to name commissioners to work with Mexican representatives in the study of the equitable division of waters of the Rio Grande and the Colorado. The main consideration of the United States in concluding the treaty was that, if no upper limit were set on Mexico's use of water by treaty, Mexico would extend its uses in advance of the border. The treaty, eventually concluded in 1944, seemed to imply that the United States had given up its earlier claim to possess absolute sovereignty over water within its territory. Thus, Junior Secretary of State Edward B. Settinus remarked, "Each country owes to the other some obligation with respect to the water of these international streams; these mutual obligations are international in scope, not merely unilateral" (quoted in Berber 1959, 116–117).

However, on closer examination the apparently fundamental change in outlook, in recognizing customary law obligations, is somewhat ambiguous. Indeed, Settinus affirmed:

> This treaty has been brought about simply by the application of the principles of comity and equity which should govern the determination of the equitable interests of two neighboring countries in the water of international streams. (quoted in Berber 1959, 117)

This statement directly excludes the idea of a legal obligation. Berber concludes that:

> The text of the 1944 treaty, whose conclusion was very much in the United States' own interests, provides no evidence of the sacrifice of national interests in favor of the rule of customary law recognized as binding under international law. (1959, 118)

Indeed Meyers also finds, from his deep consideration of the Colorado basin settlement, that "The Mexican Water Treaty embodies an international river settlement based on considerations of prudence and national policy if not on international law" (1967, 591). Despite this, the treaty has so far worked well, and Meyers concludes, "If future problems can be approached in the same spirit of the international co-operation that prevailed in reaching a solution the omissions and ambiguities in the agreement will count for little against its positive accomplishments" (1967, 591).

Advocates of the treaty system include Berber (1959) and Chauhan (1981). Chauhan compares this method to others in the problem of international water law, as, for example, the participation of third parties or other agencies such as the ILA or ILC. He sees this participation of third parties as unsuitable because they do not have the potentiality to obtain spontaneous acceptance from the different states involved. Berber states that "Treaty-making represents the application of that highest form of political wisdom which is termed 'compromise,' a word too often used contemptuously or misleadingly" (1959, 270–271).

Other benefits include the satisfaction of direct participation by the states involved in the settlement process and the fact that, being well thought out, treaties give those involved time to understand the implications and to realize that if they have given something in terms of the treaty they have received something back in return. Thus Chauhan is emphatic when he stresses that: "The treaty, being a mutually processed and mutually accepted measure, stands the chance of being executed in true spirit to the mutual benefit of all the concerned parties" (1981, 250).

This does provide significant pointers for potential agreements in the Jordan basin.

DISCUSSION

This discussion will focus our singular attention upon the Jordan basin, although it will have a general relevance for disputes on water issues elsewhere within the troubled Middle East.

The continuing, slow, hesitant, and punctuated moves toward a deeper Israeli-Palestinian accord, with a more complete recognition of the hopes and aspirations of the various parties, could lay the basis for improved and developed relationships amongst Israel, Palestine, and the neighboring states within the Jordan basin, and thus provide the necessary conditions for developing bilateral treaties on water matters between the various co-riparians.

Let us consider the Jordan basin in three parts: the upper and the lower sections about the Jordan River itself, and a third section, covering the remainder of Israel and the presently Occupied Territories. In the north it is most unlikely that Israel would ever freely surrender control of the southern Lebanon, continuously occupied by Israeli forces since the invasion, referred to by Israel as Operation Peace for the Galilee, of June 1982, or the western Golan Heights, seized by Israel in the Six-Day War of June 1967 (Rowley 1984, 142; 1989). It is our considered belief that Israel will continue effectively to control the upper Jordan waters and proceed to implement various diversionary schemes for waters within this crucially important sector (see also Amery 1993; 1997).

The situation within the second sector, the middle and southern parts of the Jordan basin, is considered to be much more hopeful, and the likelihood is that here there will be a compromise agreement on water issues between Israel and the Hashemite Kingdom.

One move toward such a compromise may lie within the Israel-Jordan Peace Treaty of 1994, which includes a section on "Water Related Matters" (see ILM 1995). While Nurit Kliot (in this volume) provides a detailed documentation relating to water in both the Israel-Jordan Peace Treaty and the Israeli-Palestinian Agreement on Water and Sewage, several points are notable. The Joint Water Committee, referred to in the above formal treaty, is again only to possess permissive powers, and the section on "Water Related Matters" (ILM 1995, 58–60) is open to various interpretations, so as to provide no hard and fast enforceable regulatory function or commitment. Again, this section is replete with the use of "may" rather than "will" (see ILM 1995). Nevertheless, the Israel-Jordan Peace Treaty of 1994, particularly Annex II (see Kliot, this volume), as it relates to water issues, representing both an allocation and cooperative agreement and also a joint management protocol, indicates the level of recent cooperation. It also serves to reaffirm the view expressed earlier concerning the importance of treaty-making.

The U.S.-brokered Israeli-Jordanian treaty can also be considered as a further step in the broader U.S. plan for the region: to establish bilateral treaties between Israel and each of its Arab neighbors. The payoff for Jordan from the United States includes at least $700 million in debt relief, together with new military aid, various U.S. aid and collateral for joint tourist ventures, a $1 billion Dead Sea development program, and better communications, including road and bridge border crossings and deep water port facilities.

One notable "peace dividend" could also see an increased cooperation between Israel and the Hashemite Kingdom of Jordan on larger-scale water issues. For example, Pearce (1991) considers the Hashemite

Kingdom's proposal for linking the Dead and Red Seas, by constructing a canal along the Wadi el Jeib, to the Gulf of Aqaba in the deep south. Such a scheme, apparently a far cheaper option than the earlier Israeli plan for a tunnel and canal from the Mediterranean to the Dead Sea (see Figure 9.1), both in construction and operating costs, would serve to restore the current low level of the Dead Sea. It would also provide hydroelectricity to power desalination plants.

While certain attendant problems would result, Pearce (1991) emphasizes the significant positive aspects of the Aqaba–Dead Sea canal project, which would derive from a progressive peace process. Wolf and Murakami (1995) have also considered the technical, environmental, economic, and political factors to establish the overall viability of regional water development projects for the Dead Sea and Aqaba, while Murakami and Wolf (1995) examine the water-energy development alternatives in the same region, the shared resources and wider benefits that would follow from a possible multilateral peace agreement amongst Israel, Palestine, and Jordan. The plans for the Mediterranean-Dead Sea Conduit Scheme and the Aqaba Sea Water Pumped-Storage Scheme are also considered cost-effective options.

Thus, in the fullness of time, an Israeli compact with the Hashemite Kingdom could probably include firmer agreement over the Yarmuk, the major east bank Jordan tributary. Such a definite and clear rapprochement between Israel and Jordan would also permit cooperation in the Aqaba–Dead Sea canal development, and lead on to considerable economic and tourism developments in the south of both Israel and the Hashemite Kingdom.

Thirdly and finally, Israel continues in its moves to fragmentize the Occupied Territories, into what certain commentators have referred to as a "Bantuization," with the Nablus-Jenin block, the Ramallah block, the Jericho block, the Hebron block, and the Gaza Strip being separated by Israeli-controlled buffer zones (Figure 9.4).

Furthermore, the decision by Israel, announced on 30 November 1996, that it would expand Jewish settlements in the West Bank was followed on 1 December by Prime Minister Benjamin Netanyahu's statement that "Israel will keep the Jordan Valley in the West Bank as Israel forever," and that "the Israeli government considers the Jordan Valley an inseparable part of Israel in any permanent settlement of the region" (Miller 1996, 7). Likewise, the announcement by the Israeli government in mid-December 1996 that it "would restore financial incentives to people [settlers] willing to live in the West Bank and Gaza" (Genkin

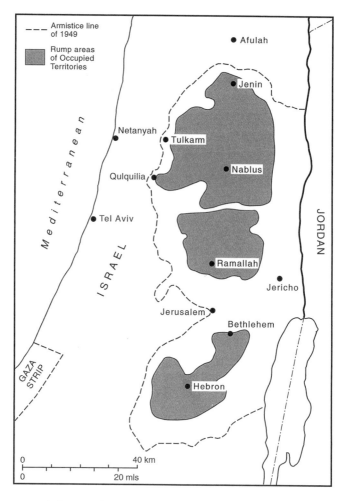

FIGURE 9.4. A future Palestinian territory? Redrawn from the Israeli newspaper *Ha'aretz,* 7 May 1992.

1996) provides a further ominous pointer to deepening Israeli control of much of the Occupied Territories. Such a view is reinforced by Collins (1996), Finkelstein (1995), and Chomsky (1996) in their reviews of Oslo II, the interim agreement signed by the late Israeli Prime Minister Yitzhak Rabin and PLO Chairman Yasser Arafat in Washington on 28 September 1995. For Collins, Oslo II is best characterized "as a document that guarantees the permanent presence of a dominant Israeli entity in the West Bank while the Palestinians are confined to the enclaves, the ultimate boundaries of which are to be effectively determined by Israel" (1996, 17).

Thus within this morcellated structure, as depicted in outline within Figure 9.4, Israel would be set to retain firm and complete control over water resources. As Meron Benvenisti et al. succinctly assert: "Control over West Bank water must remain in Israel's hands in all circumstances" (Benvenisti, Abu-Zayed, and Rubinstein 1986, 223).

On this issue Israel would brook no compromise, and the Israeli water authority, Mekorot, is working on the complete integration of the entire West Bank system into the larger Israeli regional network. Indeed, even with continued control of the upper Jordan and the West Bank aquifer, Israel still faces a water deficit situation, and in most years Israel's water consumption exceeds the renewal rate of supplies.

Despite such larger-scale initiatives a number of researchers are greatly concerned with longer-term water deficits and conservation issues. Barberis (1991) reports upon the sensitive matters relating to the development of régimes for the maintenance and protection of shared aquifers, while Arlosoroff (1995) considers the water supply in Israel, Jordan, and an envisaged Palestinian state. Arlosoroff reviews the main techniques to conserve water, such as effluent reuse and water-efficient irrigation, and how such methods would keep pace with the social and food-production needs of an anticipated 40 percent growth in population within the region by the end of the century. Alternatively, Stauffer proposes a dramatic reduction in Israeli agriculture and hence water usage. "Basically, two-thirds of [Israeli] agriculture enjoys such large [financial] subsidies, that, by reducing water use, thereby reducing subsidies, the net [GDP] benefit is positive" (1996, 14). Such a reduction is, however, highly improbable.

Clearly, even a complete Israeli control of all Jordan basin waters would, in the longer term, be insufficient to resolve such emergent problems. A new sense of real compromise must now emerge amongst the various co-riparians within the Jordan basin. As an example of such a new period of détente, Josephs (1996), for example, considers a series of massive development proposals, together costed at in excess of $30 billion, that could transform southern Israel, Gaza, and neighboring Arab countries, which was put forward by Israel at the 1994 Middle East Summit in Casablanca.

Yet even with such developments as the Dead Sea replenishment scheme Israel will still find itself in a water deficit situation that may require a long-distance importation of water overland from outside the region, probably from Turkey. Such a massive undertaking would probably need the acquiescence and support of both Syria and the Lebanon, and thus necessitate a longer-term compromise upon upper Jordan waters with these co-riparians, as opposed to the current Israeli power politics in the upper Jordan, the south Lebanon, and the Golan.

Finally, in this discussion it is of interest to record that, when I first became a professional geographer back in the mid-1960s, a particular academic debate centered on the possibilist-determinist controversy (see Taylor 1993, 146–147). The creation of Israel and its subsequent colossal growth and development into a modern industrial state, with a gross national product of some $85.5 billion, a population of over 6 million, and an average per capita income of about $18,000 a year, in the dominantly semiarid conditions about the eastern margins of the Mediterranean, are now having profound economic and environmental consequences upon the entire region and lend some support to the environmental-determinists' base argument of particular environmental constraints.

CONCLUDING REMARKS

This chapter has considered the general lack of controls on river waters and their abstractions within the Middle East in general and the Jordan basin in particular. Many of the raging conflicts within the Middle East can be related to water issues, and it does seem quite incredible that few formal laws have been formulated and implemented, as on water use within international river basins. The problem is that, as quite briefly demonstrated in this chapter, the formulation of general, hard and fast legal relationships is not usually feasible, especially when each and every river basin is different and where the relationships between the co-riparian states also differ.

Indeed, it has been suggested that "A dangerous source of error and inadmissible generalization in international law is the belief that all international relationships constitute legal relationships, and that it is possible to remove all disputes and grievances from the world by the application of rules of law" (Berber 1959, 121).

Clearly, that is not the case. Therefore, it is concluded that separate treaties between the various states, incorporating specific compromises, would provide a potential basis for a resolution to the problems. Indeed Chauhan stresses that "A dispute between states can be settled only through the position of compromise and the spirit of give and take" (1981, 200).

However, even this may not be enough on its own, and the co-riparian states will need also to recognize the interdependence and finite limitations of the natural river basin, and therefore the importance for each state to work with the others. However, here "Much more work needs to be done to convince states that it is in their broader self-interests to view water as part of a system governed by the laws of nature, rather than as an aspect of territorial sovereignty" (McCaffrey 1991a, 165). The longer-term problems of competitions for water within the Middle East cannot be indefinitely attended to by a resort to military force and power plays.

Relating back to our prime concerns with the Jordan basin, it has to be realized that we do live in particularly interesting times where, on the one hand, the possibilities of peaceful coexistence and major development potentials beckon, while, on the other hand, the continued and forceful domination and control of the Jordan basin by the State of Israel point to longer-term acrimony and major conflict.

On the general situation in the Middle East overall, Tarasofsky (1993) likewise concludes that "optimal utilization" of waters can only really come about with "comprehensive communal management." He sees that a strong institutional structure in this area is needed, one that is capable of taking decisions to solve the region's needs, particularly in the Jordan basin, where he, too, predicts that water will eventually have to be imported.

It must also be forcefully emphasized here that I fully expect there are many who will criticize this present presentation for introducing and discussing certain of the issues raised, for adopting what I prefer to call an overtly realistic rather than a too optimistic position, and in noting some of the considerable difficulties that still lie ahead. Such viewpoints derive directly from my proceeding programs of interative fieldwork (Rowley 1999).

Let us pray that reason will finally prevail and that realistic compromises over water issues may be reached to enhance future development and provide a real sense of hope for all within this most sensitive region.

ACKNOWLEDGMENT

I thank Dr. Rick Cryer and Dr. David Knighton for their help with the hydrological data.

NOTES

1. Smith (1966, 119–122) provides outline details of a number of schemes, including the Baker-Harza Plan, and the earlier Hays-Savage and Lowdermilk schemes, together with the later Israeli Seven-Year Plan of 1953, the Cotton Plan

of 1954, and the Israeli Ten-Year Plan of 1956 (see also Amery and Kubursi 1994, 184–185).

2. The Israeli "annexation" of the Golan Heights is considered by the Israel Defense Forces Spokesman as follows: "The decision of the Israeli Government (14/12/81) to extend Israel's 'law, jurisdiction and administration' to the Golan Heights nullifies the Israel Military Administration. Israel acted as a sovereign nation to bring civilian law to an area which had been under military regulations for the past 14 years.

"The Israeli Government has therefore also assumed direct administration of the health, welfare, agriculture, employment and education of the residents of the Golan Heights—thus relieving the IDF [Israel Defense Forces] of all other functions except the defense of the area against hostile activity" (1981b, 37).

In essence, it would appear the Israeli position is that the Golan Heights is now a part of the State of Israel.

REFERENCES

Allan, J. A. 1992. "The Changing Geography of the Lower Nile: Egypt and the Sudan as Riparian States." In G. P. Chapman and K. M. Baker (eds.), *The Changing Geography of Africa and the Middle East*, pp. 165–190. London: Routledge.

Amery, H. A. 1993. "The Litani River of Lebanon." *Geographical Review* 83 (3): 229–237.

———. 1997. "Water Security as a Factor in Arab-Israeli Wars and Emerging Peace." *Studies in Conflict and Terrorism* 20 (1): 95–104.

———, and A. A. Kubursi. 1994. "The Litani River: The Case against Inter-basin Transfer." In D. Collings (ed.), *Peace for Lebanon? From War to Reconstruction*, pp. 179–194. Boulder, Colo.: Lynne Reinner.

Anderson, E. 1991a. "The Violence of Thirst." *Geographical Magazine* 67 (2): 31–34.

———. 1991b. "White Oil" *Geographical Magazine* 67 (5): 10–14.

Arlosoroff, A. 1995. "Promoting Water Resource Management in the Middle East." *International Water and Irrigation Review* 15 (2): 6–12, 15–16.

Barberis, J. 1991. "The Development of International Law of Transboundary Groundwater." *Natural Resources Journal* 31 (1): 167–186.

Benvenisti, M., Z. Abu-Zayed, and D. Rubinstein. 1986. *The West Bank Handbook: A Political Lexicon.* Jerusalem: Jerusalem Post.

Berber, F. J. 1959. *Rivers in International Law.* London: Stevens.

Chauhan, B. R. 1981. *Settlement of International Water Law Disputes in International Drainage Basins.* Berlin: Schmidt.

Chomsky, N. 1996. "Eastern Exposure: Misrepresenting the Peace Process." *Village Voice*, 6 February, pp. 6–8.

Collins, F. 1996. "How Oslo II Carves Up the West Bank." *Washington Report on Middle East Affairs*, December, pp. 17, 89.

Elliot, M. 1991. "Water Wars." *Geographical Magazine* 63 (1): 28–30.

Farnham, H. R. 1904. *The Law of Water and Water Rights: International, National, State, Municipal and Individual*, 3 vols. Rochester, N.Y.: Lawyers Co-op.

Finkelstein, N. G. 1995. *Image and Reality of the Israel-Palestine Conflict*. New York: Verso.

Genkin, S. 1996. "Row Escalates over Jerusalem Housing." *Jewish Chronicle*, 20 December.

Hudson, H. 1970. "The Litani River of Lebanon: An Example of Middle Eastern Water Development." *Middle East Journal* 25 (1): 1–14.

ILC. 1994. *Draft Articles on the Law of the Non-Navigational Uses of International Watercourses. Draft Report of the International Law Commission. UN.GAOR. 43rd Session. UN. Doc. A/CN 4/L 463/Add.4 (1991)*. Geneva: United Nations.

ILM. 1995. "Israel-Jordan Treaty of Peace (Done at the Arava/Araba Crossing Point, October 26, 1994)." *International Legal Materials* 34 (January): 43–66.

Irani, R. 1991. "Water Wars." *New Statesman and Society* 3 (1): 24–25.

Israel Defense Forces Spokesman. 1981a. *The Lebanese Border*. Jerusalem: Israel Defense Ministry.

———. 1981b. *The Golan Heights*. Jerusalem: Israel Defense Ministry.

Jaber, R. 1989. "The Politics of Water in South Lebanon." *Race and Class* 31 (1): 63–67.

Josephs, B. 1996. "Massive Mideast Projects Could Transform Region." *Jewish Chronicle*, 13 April.

Kahhaleh, S. 1981. *The Water Problem in Israel and Its Repercussions on the Arab-Israeli Conflict*. Beirut: Institute for Palestinian Studies.

McCaffrey, S. C. 1991a. "International Organizations and the Holistic Approach to Water Problems." *Natural Resources Journal* 31 (2): 139–165.

———. 1991b. "United Nations International Law Commission Report on the Draft Articles Adopted at Its Forty-Third Session." *International Legal Materials* 30 (6): 1554–1562.

Macrory, R. 1985. *Water Law: Principles and Practice*. London: Longman.

Meyers, C. J. 1967. "The Colorado Basin." In A. H. Garretson (ed.), *The Law of International Drainage Basins*, pp. 486–608. New York: Oceana.

Miller, M. 1996. "Israel OKs More Jewish Housing in the West Bank." *Los Angeles Times*, 2 December.

Murakami, M., and A. T. Wolf. 1995. "Techno-political Water and Energy Development Alternatives in the Dead Sea and Aqaba Regions." *International Journal of Water Resources Development* 11 (2): 163–183.

Newman, D. 1994. "Towards Peace in the Middle East: The Formation of State Territories in Israel, the West Bank and the Gaza Strip." *Geography* 79 (2): 263–268.

Oppenheim, L. F. L. 1955. *International Law: A Treatise*, 8th ed. London: Longman-Green.

Pearce, F. 1991. "Wells of Conflict on the West Bank." *New Scientist*, 1 June, pp. 36–40.

Postel, S. 1992. *The Last Oasis: Facing Water Scarcity*. London: Earthscan.

Pugh, D. 1990. "Nile Waters." *Guardian*, 12 October.

Rowley, G. 1984. *Israel into Palestine*. London: Mansell.

———. 1989. "The Lebanon: From Conflict and Change to Cantonization?" *American Geographical Society's Focus* 39 (3): 9–16.

———. 1990a. "The Jewish Colonization of the Nablus Region: Perspectives and Continuing Developments." *GeoJournal* 21 (4): 349–362.

———. 1990b. "The West Bank: Native Water Resource Systems and Competition." *Political Geography Quarterly* 9 (1): 39–52.

———. 1990c. "On Human Dominion over Nature in the Hebrew Bible." *Annals of the Association of American Geographers* 80 (3): 447–451.

———. 1992. "Human Space-Territoriality-Conflict. An Exploratory Study with Special Reference to Israel and the West Bank." *Canadian Geographer* 36 (3): 210–221.

———. 1993. "Multinational and National Competition for Water in the Middle East: Towards the Deepening Crisis." *Journal of Environmental Management* 39 (2): 187–197.

———. 1995. "Review of Ruth Lapidoth and Moshe Hirsch (eds.), *The Jerusalem Question and Its Resolution*." *GeoJournal* 37 (1): 190–191.

———. 1999. "'Down to Earth' within the Occupied Territories" (in preparation).

Smith, C. G. 1966. "The Disputed Waters of the Jordan." *Transactions of the Institute of British Geographers* 40 (1): 111–129.

Smith, H. A. 1931. *The Economic Uses of International Rivers*. London: King.

Stauffer, T. R. 1996. *Water and War in the Middle East: The Hydraulic Parameters of Conflict*. Washington, D.C.: Center for Policy Analysis on Palestine.

Tarasofsky, R. G. 1993. "International Law and Water Conflicts in the Middle East." *Environmental Policy and Law* 23 (1): 70–73.

Taylor, P. J. 1993. *Political Geography: World-Economy, Nation-State and Locality*, 3d ed. London: Longman.

Teclaff, L. A. 1991. "Fiat or Custom: The Checkered Development of International Water Law." *Natural Resources Journal* 31 (1): 45–73.

Thornthwaite, C. W., J. R. Mather, and D. B. Carter. 1958. "Three Water Balance Maps of South-West Asia." *Publications of the Drexel Institute of Technology* 10 (1): 57–59.

Todd, D. 1992. "Common Resources, Private Rights and Liabilities: A Case Study of Texas Groundwater Law." *Natural Resources Journal* 32 (2): 233–263.

Vesiland, P. J. 1993. "The Middle East's Critical Resource." *National Geographic* 183 (3): 38–71.

Wolf, A. T., and M. Murakami. 1995. "Techno-political Decision Making for Water Resources Development: The Jordan River Watershed." *International Journal of Water Resources Development* 11 (2): 147–162.

10. *The Spatial Attributes of Water Negotiation*
THE NEED FOR A RIVER ETHIC AND RIVER ADVOCACY IN THE MIDDLE EAST

John Kolars

As noted by Wolf, negotiations regarding the sharing of scarce water supplies usually begin with claims based upon chronology or regional hydrology. That is, prior usage establishes current rights, or upstream/source locations dictate control of transboundary rivers. Such positions seldom lead to reasonable settlements of differences in sharing upstream/downstream waters. Wolf further demonstrates that most amicable solutions to such differences are based on the needs of disputants after careful examination of how such needs can be defined (Wolf 1995).

This discussion recognizes the efficacy of such an approach but wishes to point out a new area important to negotiation and mediation: river advocacy based upon a river ethic.[1]

THE CONCEPT OF RIVERINE NEED

The concept of need is a good place to begin. Simply put, just as human beings need the river, the river needs human beings. The river needs their consideration, their help, their perception that the river is an entity that to survive must maintain a healthy corpus in a felicitous environment. If the river perishes, so will all the creatures, great and small, human, animal, and plant, with which it shares a commensalistic relationship and fate.

THE RIVER AS ENTITY

Consider the following: a river has three roles (Tenenbaum 1994, 11–15).

It serves as a routeway for the basin/region it serves, not only for human commerce, but for anadromous fish, and as a flyway for birds. Improperly used, it can also be the means of introducing disease and parasites—cholera, coli, schistosomes—along its length.

A river also serves the area through which it flows. It is the artery that provides life itself, as do its tributaries and distributaries, whether through irrigation systems or natural deltas. Not only does it collect and bring life-giving water, but it also carries away poisons and waste, serving the body regional as does a healthy circulatory system cleanse the human body.

A river is an entity unto itself. A river in its natural state assumes a comprehensive gradient from source to mouth. Disturbance or blockage of this gradient telegraphs itself from beginning to end and initiates a series of readjustments in slope and flow and carrying capacity throughout the system (Stolum 1996). A river adjusts to variations in seasonal changes in volumetric flow by means of its deepest channel (thalweg), its floodplains, and its underground flow. In its natural state, a river neither floods disastrously, nor shrinks capriciously. Those crises are defined by human perceptions, and often brought about by human intervention. For example, the increasingly destructive floods in the American Midwest are the result of inept manipulation of the region's river patterns and topology, and increased populations rushing headlong to occupy untenable floodplains.

The river has its own regulatory adjustments and its own dependent biota. All of these can be disturbed, even destroyed, through human intervention. When the river subsequently sickens, perhaps dies, the symbiotic benefits accruing to humans from the relationship cease.

THE RIVER IN SPACE: PUNCTIFORM, LINEAR, AREAL, AND VOLUMETRIC VIEWS

Why is all this not immediately evident to river users and managers and negotiators? It is because, like the blind men and the elephant, each and every one of us who use and deal with a river view it as a series of isolated parts, not as a whole. We are led to such misconceptions by our perception, conjecturing, and legal defining of rivers in either punctiform, linear, areal, or volumetric terms removed from holistic reality.

It is at this juncture that we should turn to the seminal work of geographers in order to understand the nature of such an approach. It has been demonstrated that all geographic and topological thought must begin with the above four basic spatial concepts (Nystuen 1963). This approach intersects with the work of geographers regarding human perceptions of their physical environment and the emotional responses which are thus evoked (Tuan 1974). Studies of cognitive mapping (i.e., mental maps) show, in turn, how pervasive and persuasive such views of the world can be (Thrower 1972). What is suggested here is that fundamental spatial codes and perceptions of the environment and its resources direct, bias, and guide human resource management. The treatment and utilization of rivers offer explicit examples of this phenomenon.

The most simplistic approach is when a river is viewed solely in terms of its source, if indeed it does rise from a single spring or lake. All else, the entire length of the river, may be treated by those who control such sources as inconsequential. "It is ours," they say, "to control like any other piece of chattel." *But water is a moving resource,* and cannot be subject to the ordinary laws governing immobile resources.[2] Its source may seem to be a point, but that point in turn derives from other flows, be they subterranean or atmospheric. *The source is only a node in a multidimensional system, one that cannot be owned or controlled in its entirety.*

If we see the river in linear terms, it becomes a line to cut up like a sausage, so much for me, so much for you; to dam and divert. If my share is longer, I win. The consequences of this approach can be socially as well as environmentally destructive. In the words of Ilter Turan, "The tendency of each riparian on a river to exploit its part of the river, without significantly taking into consideration the concerns of the others, produces an environment in which all riparians are suspicious of each other. This mutual lack of trust stands in the way of cooperation, since each party fears that the others are ill-intentioned, will not keep their word, exploit one's weaknesses, etc." (1996, 10–11).

We forget that, as with a snake, or the spinal cord, or the aorta, to cut the river into sections is to destroy it. Its integrity depends upon its unity. An ancient story illustrates this point. King Solomon was approached one day by two women, each fiercely clutching the same child by an arm. Each woman claimed that the child was hers and hers alone. They asked Solomon to judge which woman should have possession of the child. The evidence was confusing and contradictory. Solomon, in his wisdom, ordered that the child be cut in half and one part given to each of the two claimants. The first woman accepted his judgment; the second said, "Give the child to the other woman; I would not have it injured, no matter how much I want it." The king awarded the child to the second woman whose love indicated her true maternal instincts. Unless we, in our management of water resources, recognize that the river is like the child and exercise the wisdom of Solomon, everyone will suffer, not only the river.[3]

Similarly, we seldom think of where the river ends. Our thoughts are always on its sources and who controls them. But what of its debouchment and where the wastes the river carries finally accumulate? Or of the waters which it may feed and sustain? What of the ecologies of lakes and playas and gulfs and seas? Our linear thinking is too often asymmetrical and selfish. There are numerous examples of this, including the impact of the use of the Danube River on the Black Sea (Pringle et al. 1993), and my prediction of similar possible repercussions of upstream developments along the Euphrates and Tigris Rivers on the headwaters of the Arabian Gulf (Kolars 1994a).

Areal thinking usually takes the form of a contest between potential sharers, each trying to list the largest amount of land to be irrigated. What follows is a grand siphoning off of every possible drop of water by each party. The result is a river unable to cleanse itself, unable to carry off its load of natural and human waste. As it is not the cholera bacillus which kills us, but the unexpurgated poisons resulting from the death of millions of bacteria, so we foul our own abodes. All this results from a reflex based on fear and greed, fear of being cheated, and that all too human constant, wanting more than we really need. This in turn leads to questions of water security and food security, rendered meaningless when careful thought reveals that we are always in a state of insecurity, always dependent upon someone else for our food.

Returning for a moment to the linear way of thinking, research has shown that occupants along a river view it as a source of clean water upstream and as a conduit for waste downstream. Again, segmental thinking obviates the true nature of the river.

Finally, volumetric perceptions of the river lead us inevitably into fixed-share thinking. When we are presented with figures stated in volumes, that is, cubic meters per year, we imagine neat stockpiles of water aggregated like so many bales of hay to be divided up among consumers. If there are so many millions of cubic meters of water annually, or so many cubic meters of flow per second, then each one of us must have as large a share as possible. Each cites the greatest needs, urban, domestic, industrial, and let the devil take the hindmost, including the river.

Thinking only of volumes of river water overlooks the dynamic of the river itself. There are times of surplus and times of low flow. There are seasonal variation and multiyear variation, and both are largely unpredictable. The proper impounding of water for future use can do much to eliminate such uncertainty, but when annual flow data are assumed to be volumes fixed in stone, then downstream sharers expect and demand unreasonable amounts of water in years of low flow (see Table 10.1).[4]

TABLE 10.1. Average Yearly Flow (cms) of the Euphrates River: 1937–1963 — (a period of generally low flows)

Year	cms	Downstream	Short	Long	Cumulative	Assumed Storage bcm (Turkey keeps 450/yr)	
		assumed full storage at beginning of series in cms				+1484	
37	894	444	−56		−56	1428	
38	997	547		+47	−9	1475	
39	831	381	−119		−128	1356	
40	1165	715		+215	+87	1484*	
41	1120	670		+170	+257	1484*	
42	1032	582		+82	+339	1484*	
43	856	406	−94		+245	1390	
44	1056	606		+106	+351	1484*	
45	691	241	−259		+92	1225	
46	920	470	−30		+62	1195	
47	703	253	−247		−185	948	
48	1007	557		+57	−128	1005	
49	662	212	−288		−416	717	
50	753	303	−197		−613	520	
51	716	266	−234		−847	286	
52	932	482	−18		−865	268	
53	906	456	−44		−909	224	
54	1012	562		+62	−847	286	
.... reservoirs empty beyond this point							
55	588	138	−362		−1209	0	−76
56	827	377	−123		−1332	0	−199
57	818	368	−132		−1464	0	−331
58	655	205	−295		−1759	0	−626
59	574	124	−376		−2135	0	−1002
60	826	376	−124		−2259	0	−1126
61	484	34	−466		−2725	0	−1592
62	692	242	−258		−2983	0	−1850
63	1356	906		+406	−2577	0	−1444

(27-year average flow = 854.6 cms = 27.0 × $10^9 m^3$/yr)
(Assumed storage 46.8 bcm = 1488 cms) (Turkey uses 450 cms/yr)
(*Short* and *Long* refer to values above and below 500 cms downstream)
* Additional surplus lost—assuming no additional storage.

They forget that the river is a varying entity. Everyone simply wants his or her own share of the stated volume.

But such thinking conceals the fact that the data are multiyear averages and that the flow of the river not only varies from year to year but that such averages are often inaccurate or simply wrong. Thus, river sharers end up playing the "sheep on the steppe" game. If a head tax is planned, there are very few sheep indeed; if a fodder bonus per head is offered, a miraculous number of sheep are presented to be counted. Upstream paucity and downstream need become inevitable.

How can all this be avoided? We must think of the river as a total entity with an existence of its own. We must realize that our relationship to the river is a truly symbiotic one. We must learn to think holistically, not in terms of points, lines, areas, and volumes. We must extend the time frame of our perceptions to spans of years, decades, to future droughts and surpluses and burgeoning populations. We must see the river as a living entity and become its advocates. We must subscribe to a river ethic.

How can these ideas be translated successfully into a realpolitik, in order to facilitate give and take across the mediation table? The participants must recognize the necessity of preserving the river's rights as well as their own, that they share an undeniable mutual dependency with the river. Let the river be seen as a vulnerable whole, the concern of all who sit at the mediation table. If the problems faced by the river are considered first, it may become possible to finesse seemingly irreconcilable differences between riparians.

RIVER ADVOCACY: AN APPLICATION

An hypothetical illustration of such river advocacy is presented in the section which follows. There are three elements underlying this example. The first is the postulated necessity of future water imports to the Jordan River basin. Such imports are suggested on the basis of a recent analysis of the situation described below (Middle East Water Commission). The second element is that the source of such imported water could be Turkey, and that such imports would be by pipeline across Syria. The third is that before such trans-Syrian water movements could take place, the question of the sharing of the Euphrates River by Turkey, Syria, and Iraq would have to be equitably resolved. It is at this juncture that the role of river advocacy, that is, representation of the rights of the Euphrates River itself, might help resolve seemingly intractable problems among the three riparians.

JORDAN BASIN WATER NEEDS

The comments in this section refer to the work of the Middle East Water Commission (MEWC) of the International Water Resources Association (IWRA). Others have discussed the continuing and growing need for water in the Middle East and have painted a broad picture of the priorities involved (Rogers 1994). Such efforts describe in general terms the problems faced in satisfying future thirst and their possible solutions. The purpose of the MEWC was to suggest a more specific blueprint for the hydrodevelopment of the Jordan River basin and the area which surrounds it. The authors also tried to extend the planning horizon far enough into the future to consider the possibility of importing water from outside the region, the most likely source of which would be Turkey.

The MEWC recognized three major parameters affecting conditions in the Jordan basin: (1) The area experiences high seasonal and multi-annual variance in precipitation and attendant stream flow. (2) The area lies between a region of rain-fed agriculture to the north and arid regions dependent upon irrigation to the south, and as a transitional zone requires a mix of technologies common to both. (3) Unchecked population growth will within thirty years preclude any use of indigenous fresh waters save for domestic purposes.

The major conclusions and suggestions made by the MEWC emphasize the necessity of a combination of short-term, medium-term, and long-term responses to growing water shortages. Technical responses included the combined production of hydroelectricity (HE) *and* reverse osmosis (RO) water, and the use of off-peak electricity for pumped storage to meet peak-demand electric generation. This, in turn, would necessitate the careful balancing of the level of the Dead Sea in order to accommodate introduction of either Mediterranean or Red Sea water for HE and RO production as well as future inputs of northern waters. As needs in the Jordan basin increase, the importing of water appears necessary—possibly from Turkey—by Medusa-type bags to coastal areas and/or pipelines to inland locations in Jordan and Palestine, and to the cities of Syria.

When considering importation by pipeline, the need becomes apparent that long-range planning should begin as soon as possible. By the same token, emphasis on cooperative regional development and a search for solutions to parallel problems among the riparians on the Euphrates is critical. Syria cannot be expected to sanction the movement of water across its territory to other nations until it feels its own needs have been equitably met. Regard for the good of the river by all its riparians may provide the common ground upon which they can first agree. It is this issue which the following discussion of the role of river advocacy addresses.

THE SOUTHEASTERN ANATOLIA DEVELOPMENT PROJECT AND THE CONUNDRUM OF THE EUPHRATES RIVER

The idea of importing water into the Jordan basin from either the Euphrates or elsewhere in Turkey is not new. Sir Hamilton Gibbs was perhaps the first to suggest (in 1946) a pipeline from the Euphrates in Iraq to Amman. The best-known scheme is former Turkish President Turgut Özal's Peace Pipeline, which he proposed should run from the Seyhan and Ceyhan Rivers to the Arabian Peninsula as far as Sharjah in the east and Jiddah in the west (Brown and Root 1987). Subsequent suggestions included Kolars's Mini-pipeline to Amman (Kolars 1992) and Wachtel's Peace Canal to the Golan Heights (Wachtel 1995).

All such schemes depend upon the availability of water in Turkey, the willingness of the Turks to part with such water, the compliance of Syrians in allowing water to pass to the south through their country, and the resolution of the question of sharing the waters of the Euphrates by Turkey, Syria, and Iraq. Answers to these questions depend in turn upon a realistic view of the amount of water the Euphrates can provide. However, no riparian participant, to date, has suggested that the river itself should be protected in order to preserve the ecological viability of the region.

The projected use of the river by the three riparians could exceed the actual flow of the river in the year 2040 by at least 2.0 bcm/yr (Özal and Altinbilek 1994, Table 8; see also Kolars and Mitchell 1991, Figure 11.1, for a similar prediction). The flow of the Tigris River would be reduced by almost 50 percent, with another 2.0 bcm removed to meet the above shortfall. It should be noted that the 2.0 bcm in question would not restore natural flow to the Euphrates but merely meet agricultural shortages. The Euphrates River would be dry to its confluence with the Tigris at Qurna. Such a situation is ecologically unacceptable and raises significant questions regarding the impact of such conditions upon the Arabian Gulf and the Gulf Cooperation Council (GCC) nations (Kolars 1994a).

Turkey has been accused by Syria and a coalition of Arab nations including Egypt and the GCC states of denying sufficient downstream flow to Syria and Iraq, and of polluting the water it does release downstream (*Turkish Daily News* 1996).[5] That such an accusation is essentially political in nature is shown by an examination of the actual situation. Turkey has on three brief occasions limited the downstream flow of water for technical reasons relating to the construction of dams within the Southeast Anatolia Development Project (GAP). Such events have been temporary, and the major one involving work on the Ataturk Dam spillways was announced in advance and compensated for by Turkey's increasing downstream discharge prior to the event.

True, the future may bring shortages to the river, but these would be not by Turkish actions alone but by those of all three riparians. In fact, Turkey, of the three, has shown the greatest awareness of the problems involved and of possible solutions to them (Bilen 1994b).

By the same token, present claims of pollution from Turkish sources cannot be accurate, for no major upstream agricultural or industrial activities are in place and operating which could account for such a situation. The only place where pollution may be occurring at present would be in runoff from the Harran Plain south of 'anliUrfa, where uncontrolled pumping of groundwater has resulted in rapid expansion of irrigated cotton agriculture *before* the arrival of Euphrates waters from Lake Ataturk via the Urfa tunnels (personal observation by this author). Such increased production is indicated by recent analyses of satellite imagery (Wannebo 1995; Beaumont 1996). Lavish overuse of fertilizers and pesticides in the area could result in the pollution of the Balik and Khabur Rivers, which drain south into Syria. If such a situation does exist, it should be confirmed and corrected.

All such disagreement is underlain by the legal positions held by each of the riparians. Iraq cites the right of prior usage, dating its claim to both the recent and very distant historic past. Turkey claims the right of sovereignty, for 88 percent of the flow of the main stream of the Euphrates originates in Turkey and another 10 percent from the Balik and Khabur derives from aquifers charged north of the Turkish-Syrian border. Syria claims both its need for agricultural and domestic water and the principle of equity.

All three riparians appeal to the latter principle, though in somewhat different ways. Iraq and Syria have claimed 750 cms (approximately 75 percent of average flow) and have also stated that the flow should be divided equally, one-third to each nation. (The latter two countries have also agreed to share whatever water crosses the border in the main stream from Turkey into Syria, with 58 percent going to Iraq and 42 percent to Syria.) Turkey has countered with a proposal that a complete survey of soils and crops in the river basin should be conducted by all three countries in consort. Thereafter, equitable need would be defined through the rational application of water to the appropriate crops where needed (see Republic of Turkey 1995 for a discussion of these issues).

Needless to say, other matters cloud the situation. Terrorism by the Kurdish Socialist Workers Party (KPP), questions of KPP sponsorship and its geographical origins, the Turkish claim of the overuse of the Orontes (Asi) River by Syria and the shorting of flow into Hatay Province, and a plethora of simmering political misunderstandings discourage easy negotiations regarding use of the Euphrates.

THE ROLE OF RETURN FLOW MANAGEMENT AS A FUNCTION OF RIVER ADVOCACY

Considering the above, is there any way to decide who shall use what amount of the Euphrates River? It was in the face of such complications that this discussion has referred to the tale of Solomon and the child. If the river is not protected, in a short time it will serve no one well. Even Turkey, upstream, will face endless distractions by irate downstream riparians.

If, however, Solomon's decision is heeded and the welfare of the river is considered, a solution can be found. For such a solution to work: (1) All the riparians must recognize that no one can claim everything each desires. There simply isn't enough to go around. (2) The river must be protected to ensure its ecological viability for the good of the region and the gulf into which the combined flow of the Euphrates and Tigris Rivers enters. The example that follows assumes that the river needs a minimum of 5 bcm uninterrupted flow per year in order to sustain itself. (3) The river must be viewed holistically rather than in terms of points, lines, areas, and volumes. (4) The riparians must accept the fact that reasonable amounts of water for each of them can be provided as long as *acceptably clean return flow (RF) is sent downstream by each of the users.*

There are numerous ways in which such a holistic view could be implemented, although space does not allow a comprehensive approach. Let us consider only one scenario that might be possible if the riparians agreed to support an advocacy role for the river.

The importance of clean RF has already been suggested (Bilen 1994a, 91). A similar solution to a parallel international problem was the building of a desalination plant on the Colorado River in Arizona to ensure acceptable downstream flow into Mexico. Given proper attention, much the same can be accomplished on the Euphrates, and both the river and its riparians can be the winners.

To suggest a general idea is one thing, to put it into practice is another. Can it be done? The remainder of this chapter describes an approach, which, while hypothetical, uses available data to demonstrate that careful management of the river and of return flow can attain desired results.

RETURN FLOW MANAGEMENT ON THE EUPHRATES RIVER — AN EXAMPLE

Table 10.2[6] illustrates the possibility suggested above. The precise management of a river with numerous tributaries and dams and multiple uses is more complex than can be presented here. This example must be, therefore, a first iteration, a hypothetical attempt at management, a beginning. Nevertheless, the reader is requested to carefully peruse the endnotes accompanying the table in order to understand its numerous steps.

In the example, progress of water downstream begins with recognition that the seasonal and multiannual high variance of flow must be controlled by upstream reservoirs, and that since all three riparians would benefit from this, all should accept as their fair share some diminution of flow caused by evaporation from the reservoirs of Turkey.[7]

Turkey, in this example, would use less river water (386 cms) than it has previously claimed (approximately 500 cms), but with careful management it would still be able to irrigate the approximately one million hectares it has anticipated. The exact proportion of return flow (RF) is somewhat conjectural, but 25 percent RF seems to be a reasonable assumption. As to water purity, Bilen states that 700 ppm pollution can be expected (1994a, 89). This is an acceptable amount as long as uncontrolled agricultural inputs are kept at a minimum (see above). In this manner Turkey would be able to send two-thirds of the river's flow downstream, thus meeting one of the downstream riparians' criteria. This amount would include the river's share, an ecological hydro-necessity, of which Turkey's portion would be 54 cms.

Next, Syria would receive the released waters from upstream. It would use its one-third share (less its portion of the river's quota, another 53 cms), while releasing 58 percent of the received flow downstream, thus meeting Iraq's agreed-upon percentage. Syria, in turn, would accept its self-imposed evaporative loss from Lake Assad. Nevertheless, careful

TABLE 10.2. Hypothetical Return Flow Management on the Euphrates River

Average flow of the Euphrates[a]	31.43 bcm/yr (996 cms)
Lost to reservoir evaporation in Turkey[b]	2.9 bcm/yr (92 cms)
Available flow (after evaporation: aev)	29.43 bcm/yr (904 cms)

5.0 bcm (160 cms) assumed necessary flow to Gulf to maintain gravity flow irrigation, surge capacity & basic riverine ecology

All quantities in cubic meters per sec (cms)

Flow after evaporation	(aev)	904		
Turkey uses		-386^c		386
Turkey releases		518	$-$ 53	River share
RF		$+$ 84	332	$RF^d = 84$
(66.6% total aev[e])		602		
River share		$+$ 54		1.050×10^6 ha[f]
Turkey sends		656		@ 1 m^3 H$_2$O/m^2
Syria receives		602 + 54		
Syria uses (33.3% total aev)[g]		-301		301
Syria releases		301	$-$ 53	River share
Return flow		$+$ 50[h]		248
(58% total from Turkey—riv. share)[i]		351	$-$ 48	Res. evap.[j]
(River share from Turkey)		$+$ 53		200 RF = 50
(River share from Syria)		$+$ 53		
Syria sends (351 + 106 riv. share)		457		420,000 ha[k]
				@ 1.5 m^3 H$_2$O/m^2
Iraq receives		457 (351 = 58%)[i] (+106)		
Local evaporation & replacement[l]		$-$ 79		
		$+$ 79		
		457		
Iraq uses		-351		RF = 88[m]
				738,000 ha[n]
				@ 1.5 m^3 H$_2$O/m^2
Iraq releases: Two River shares		106		
RF		$+$ 88		
		194		
Evaporation from river shares		$-$ 34		
Flow to gulf		160 = 5.0 bcm[o]		

[a] All data as cited in Korkut Özal and Dogan Altinbilek 1994.

[b] Reservoir areas of the Keban Dam (675 km^2), the Karakaya Dam (298 km^2), and the Ataturk Dam (817 km^2), as cited in Özal and Altinbilek, Map 2. Because these reservoirs control the

variance of the river along its entire length, it is this author's opinion that their evaporative loss should be shared equally among the riparians.

[c] In the nonbinding agreement in place during the construction of the Ataturk Dam, Turkey agreed to send 500 cms average flow downstream. Continuing this practice would mean Turkey could retain 404 cms. However, for the good of the river and in order to expedite a management agreement, Turkey in this case would use only 386 cms.

[d] Return flow is estimated at 25% of the total used. Direct observation would have to refine such an estimate, with consequent adjustments.

[e] This amount corresponds to a principle cited by the downstream riparians that the flow of the river should be divided equally into thirds. This author assumes that the actual amount should be less evaporative losses mentioned in note b.

[f] 386 cms less 54 cms for the river's share = 332 cms = 10.5 bcm/yr suitable for irrigating 1.050 million ha @ 1.0 m^3 water/m^2. This is close to the total amount of irrigated land anticipated on the Turkish Euphrates.

[g] This corresponds to the principle cited in note e.

[h] Twenty-five percent of the amount removed (see note d above). This assumes all water removed remains within the Euphrates basin. The question arises regarding the possible transfer of Euphrates water to the Aleppo basin, which is closed and offers no natural return flow to the river. Pumping or other adjustments might be necessary in such a case.

[i] (F351/602 = 58%) This amount corresponds to the agreement between Syria and Iraq that Iraq should receive 58% of the water entering Syria from Turkey, but should not, according to this author, include the river's shares.

[j] Area of Lake Assad (628 km^2). Losses estimated at 1.5 bcm (48 cms).

[k] See note f. The amount of water needed per m^2 in this case is assumed to be 1.5 m^3 as the result of a more southerly location and higher temperatures.

[l] Evaporation losses from the Qadisiyah Reservoir (550 km^2) and the Habbaniyah Reservoir (400 km^2) are approximately 2.5 bcm/yr. In this case, the losses can be made up easily by diverting water from the Tigris River by means of the Thar Thar Canal or a similar canal.

[m] Iraq faces special problems of return flow which involve the use of the Third River, a giant drainage canal which empties into the Gulf behind Bubiyan Island. Special analysis would be necessary to resolve this issue.

[n] See note k.

[o] This amount is the minimum necessary to retain surge capacity for down stream gravity fed irrigation, as well as providing a suitable volume of water to maintain viability.

SOURCE: Kolars 1997.

management should allow Syria to irrigate approximately 420,000 ha along or near the river.

Iraq would next receive its one-third share of the river plus enough extra water to attain its desired quota of 58 percent of the water entering Syria from Turkey in the main stream (see above). It would accept its self-imposed evaporative losses from two major reservoirs and replenish the river from the more abundant waters of the Tigris, which it controls. The waters available to Iraq from the Euphrates are sufficient for irrigating only 738,000 ha, rather than the 1.5 million ha it has anticipated

(Özal and Altinbilek 1994, Table 7). This is partially the consequence of the "one-third of the river's flow" principle espoused by Iraq itself as well as the reality of supplies limited by nature. Fortunately, sufficient water exists in the Tigris River to compensate for this shortfall. Iraq, thereafter, would release the two upstream river-shares sent to it, and by adding its own RF downstream should be able to achieve—after natural losses— the necessary minimum flow of 150 cms (5 bcm/yr) to the Gulf.

As for pollution, Turkey's RF should be sufficiently clean; Syria's RF, though more saline, would be reasonably diluted ($^5/_{40}$ths,[8] or about 1:8); and Iraq, by judicious use of Tigris waters, could also satisfy minimum pollution standards.

Admittedly, the above program assumes cooperation, some sacrifice, and acceptance of restrictions by all the riparians. Is this possible? The alternatives would be far worse. Furthermore, members of the Black Sea Commission and the Danube Commission, once seemingly irreparably separated by the Iron Curtain, now engage in full cooperative actions with each other for the good of all (Pringle et al. 1993). A future, no less amazing, can be possible for the users of the Euphrates and Tigris Rivers and for the rivers themselves. The catalyst for such cooperation could be recognition that without it the Euphrates will die, the ecology of the Gulf subsequently damaged, and all three riparians and the nations of the GCC will suffer.

Meanwhile, basic geographic dimensional and perceptual insights help clarify the situation.

In closing, we may ask, if an approach similar to the one presented above were to result in an acceptable sharing of the Euphrates River, might it not become possible to discuss with Turkey its providing much-needed water to the peoples of the Jordan basin? And by the same token, if Syria were satisfied with and assured of its share of the Euphrates, might it not permit passage to the south by pipeline of those much-needed auxiliary waters? And if all this were possible, might not the mutually rewarding ecological viability of the river be preserved?

NOTES

Author's Note: A portion of this lecture was given at the Fifth Stockholm Water Symposium, August 1995.

1. This concept is an extension of Sandra Postel's call for the adoption of a "Water Ethic" (Postel 1992).

2. I am indebted to John Nystuen for conversations regarding this point.

3. In another way, the well-known Tragedy of the Commons can apply. "No

one will be the worse off if I add just one more irrigated row of beans. And I will be that much richer." And so I do, and so does my neighbor, and his, and the river is depleted by a thousand tiny thefts, each one seemingly harmless. It is the same with aquifers. In the Eldorado development near my hometown of Santa Fe, New Mexico, every contractor cites how small a demand on the water supply will be made by "just one more house." And yet, withdrawals from the well-field upon which the development depends have already reached their predictable reliable limits ("Hydrologist: Homes Strain Water Supply," *Albuquerque Journal North*, Thursday, 2 January 1997, p. 6).

4. Cubic meters per second (cms) is used in Table 10.1 and Table 10.2 as a more easily rendered measure of volumes.

5. A minor pollution incident on the Balikh River, which flows from Turkey into the Syrian Euphrates, was resolved at a recent "Second Track" meeting of hydrologists. The sewer system in the Turkish agricultural boom town of Akçakale, near the Syrian border, had become overloaded, with subsequent flow of raw sewage into the stream. An exchange of information between Turkish and Syrian delegates led to the situation's being successfully resolved.

6. Table 10.2 and its notes are a revised version of a similar presentation found in *Water International* (March 1997). See footnote 3.

7. I disagree with the general conclusions reached by Matthew Richardson in his study *Guneydogu Anadolu'nun Sürdürülebilir Kalkinma Potansiyeli*. GAP and its dams are a fait accompli. Retrospective angst about their impact on the environment is locking the barn door after the horse is stolen and fails to recognize the important role the dams play in controlling variance. This is not to deny that the threatened Parlak Ibis and other species must be taken into account (1995, 42), that malaria and schistosomiasis must be closely guarded against, and that wide and uncontrolled fluctuations of river levels resulting from HE discharges—as distress conservationists along the Colorado River in the Grand Canyon (Wuethrich 1995)—must be controlled.

8. That is, one-half of the water received from Turkey minus the river's share ($602/2 = 301$ cms), plus the river's share from Turkey (53 cms), plus the river's share from Syria (53 cms) equals 407 cms of presumably acceptably pure water, plus Syria's RF of polluted water (50 cms).

REFERENCES

Beaumont, Peter. 1996. "Agricultural and Environmental Changes in the Upper Euphrates Catchment of Turkey and Syria and Their Political and Economic Implications." *Applied Geography* 16 (2): 137–157.

Bilen, Özden. 1994a. "A Technical Perspective on the Euphrates-Tigris Basin." In Alli Ihsan Bagis (ed.), *Water as an Element of Cooperation and Development in the Middle East,* pp. 81–100. Ankara, Turkey: Ayna Publications and the Friedrich Naumann Foundation.

———. 1994b. "Prospects for Technical Cooperation in the Euphrates-Tigris

Basin." In Asit K. Biswas (ed.), *International Waters of the Middle East from Euphrates-Tigris to Nile*, pp. 95–116. Oxford and Bombay: Oxford University Press.

Brown and Root. 1987. *Source to Consumer*. Houston.

Kolars, John. 1992. "The Future of the Euphrates River." In Guy Le Moigne (ed.), *Country Experiences with Water Resources Management: Economic, Institutional, Technological and Environmental Issues*, pp. 135–142. World Bank Technical Publication No. 175. Washington, D.C.: World Bank.

———. 1994a. "Managing the Impact of Development: The Euphrates and Tigris Rivers and the Ecology of the Arabian Gulf—A Link in Forging Tri-Riparian Cooperation." In Ali Ihsan Bagis (ed.), *Water as an Element of Cooperation and Development in the Middle East*, pp. 129–154. Ankara, Turkey: Ayna Publications and the Friedrich Naumann Foundation.

———. 1994b. "Problems of International River Management: The Case of the Euphrates." In Asit K. Biswas (ed.), *International Waters of the Middle East from Euphrates-Tigris to Nile*, pp. 44–94. Oxford and Bombay: Oxford University Press.

———. 1997. "River Advocacy and Return Flow Management on the Euphrates/Firat River." *Water International* 22 (1): 49–53.

———, and William A. Mitchell. 1991. *The Euphrates River and the Southeast Anatolia Development Project*. Carbondale: University of Southern Illinois Press.

Middle East Water Commission (MEWC). 1997. *Core and Periphery: A Comprehensive Approach to Middle Eastern Water*. Coedited and coauthored with Asit K. Biswas, Masahiro Murakami, John Waterbury, Aaron Wolf, International Water Resources Association. New Delhi: Oxford University Press.

Nystuen, John D. 1963. "Identification of Some Fundamental Spatial Concepts." *Papers of the Michigan Academy of Science, Arts, and Letters* 48 (1963): 373–384.

Özal, Korkut, and H. Dogan Altinbilek. 1994. "Water and Land Resources Development in Southeastern Turkey." *Water in the Islamic World—An Imminent Crisis*. Eighth Islamic Academy of Sciences Conference, Khartoum, December.

Postel, Sandra. 1992. *Last Oasis*. New York: W. W. Norton & Co.

Pringle, Catherine, George Vellidis, Francis Heliotis, Dan Bandacu, and Sergie Cristofor. 1993. "Environmental Problems of the Danube Delta." *American Scientist* 81 (4): 361–365.

Richardson, Matthew. 1995. *Güneydogu Anadolu'nun Sürdürülebilir Kalkinma Potansiyeli*. Türkiye Esnaf-Sanatkar ve Kücük Sanayi Arastirma Enstitüsü, TES-AR Yayinlari No. 15. Ankara.

Rogers, Peter. 1994. "The Agenda for the Next Thirty Years." In Peter Rogers and Peter Lydon (eds.), *Water in the Arab World*, pp. 285–316. Cambridge, Mass.: Harvard University Press.

Stolum, Hans-Henrik. 1996. "River Meandering as a Self-Organization Process." *Science* 271 (22 March): 1710–1713.

Tenenbaum, David. 1994. "Rethinking the River." *Nature Conservancy*, July–August, pp. 11–15.

Thrower, J. W. 1972. *Maps & Man: An Examination of Cartography in Relation to Culture and Civilization.* Englewood Cliffs, N.J.: Prentice-Hall.

Tuan, Yi-Fu. 1974. *Topophilia: A Study of Environmental Perception, Attitudes, and Values.* Englewood Cliffs, N.J.: Prentice-Hall.

Turan, Ilter. 1996. "Water Problems: Can the United Nations Help? A Discussion with Special Reference to the Middle East." *Water International* 21 (1): 1–11.

Turkey, Republic of, Ministry of Foreign Affairs. 1995. *Water Issues between Turkey, Syria, and Iraq.* Ankara: Department of Regional and Transboundary Water.

Turkish Daily News. 1996. News note. 24 April.

Wachtel, Boaz. 1995. "The Peace Canal Plan: A New Model for the Distribution and Management of Water Resources and a Catalyst for Cooperation in the Middle East." In *Proceedings of the International Symposium on Water Resources in the Middle East: Policy and Institutional Aspects*, pp. 137–146. University of Illinois, Urbana, 24–27 October.

Wannebo, Antoinette. 1995. "Investigation of Agricultural Land-Use Changes Resulting from the Ataturk Dam in Turkey's Southeastern Provinces—Observing Earth from Space." Unpublished thesis, Yale University, New Haven, Fall 1995.

Wolf, Aaron. 1995. *Hydropolitics along the Jordan River: The Impact of Scarce Water Resources on the Arab-Israeli Conflict.* Tokyo: United Nations University Press.

Wuethrich, Bernice. 1995. "Deliberate Flood Renews Habitats." *Science* 272 (19 April): 344–345.

Jordan-Israel Peace Treaty
26 OCTOBER 1994

ARTICLE 6: WATER

With the view to achieving a comprehensive and lasting settlement of all the water problems between them:

1. The Parties agree mutually to recognise the rightful allocations of both of them in Jordan River and Yarmouk River waters and Araba/Arava ground water in accordance with the agreed acceptable principles, quantities and quality as set out in Annex II, which shall be fully respected and complied with.

2. The Parties, recognising the necessity to find a practical, just and agreed solution to their water problems and with the view that the subject of water can form the basis for the advancement of cooperation between them, jointly undertake to ensure that the management and development of their water resources do not, in any way, harm the water resources of the other Party.

3. The Parties recognise that their water resources are not sufficient to meet their needs. More water should be supplied for their use through various methods, including projects of regional and international co-operation.

4. In light of paragraph 3 of this Article, with the understanding that co-operation in water-related subjects would be to the benefit of both Parties, and will help alleviate their water shortages, and that water issues along their entire boundary must be dealt with in their totality, including the possibility of trans-boundary water transfers, the Parties agree to search for ways to alleviate water shortage and to co-operate in the following fields:

    a. development of existing and new water resources, increasing the water availability including cooperation on a regional basis as appropriate, and minimising wastage of water resources through the chain of their uses;

    b. prevention of contamination of water resources;

    c. mutual assistance in the alleviation of water shortages;

    d. transfer of information and joint research and development in water-related subjects, and review of the potentials for enhancement of water resources development and use.

5. The implementation of both Parties' undertakings under this Article is detailed in Annex II.

ANNEX II: WATER RELATED MATTERS

Pursuant to Article 6 of the Treaty, Israel and Jordan agreed on the following Articles on water related matters:

Article I: Allocation

    1. Water from the Yarmouk River

       a. Summer period—15th May to 15th October of each year. Israel pumps (12) MCM and Jordan gets the rest of the flow.

b. Winter period—16th October to 14th May of each year. Israel pumps (13) MCM and Jordan is entitled to the rest of the flow subject to provisions outlined herein below: Jordan concedes to Israel pumping an additional (20) MCM from the Yarmouk in winter in return for Israel conceding to transferring to Jordan during the summer period the quantity specified in paragraphs (2.a) below from the Jordan River.

c. In order that waste of water will be minimized, Israel and Jordan may use, downstream of point 121/Adassiya Diversion, excess flood water that is not usable and will evidently go to waste unused.

2. Water from the Jordan River

a. Summer period—15th May to 15th October of each year. In return for the additional water that Jordan concedes to Israel in winter in accordance with paragraph (1.b) above, Israel concedes to transfer to Jordan in the summer period (20) MCM from the Jordan River directly upstream from Deganya gates on the river. Jordan shall pay the operation and maintenance cost of such transfer through existing systems (not including capital cost) and shall bear the total cost of any new transmission system. A separate protocol shall regulate this transfer.

b. Winter period—16th October to 14th May of each year. Jordan is entitled to store for its use a minimum average of (20) MCM of the floods in the Jordan River south of its confluence with the Yarmouk (as outlined in Article II below). Excess floods that are not usable and that will otherwise be wasted can be utilised for the benefit of the two Parties including pumped storage off the course of the river.

c. In addition to the above, Israel is entitled to maintain its current uses of the Jordan River waters between its confluence with the Yarmouk and its confluence with Tirat Zvi/Wadi Yabis. Jordan is entitled to an annual quantity equivalent to that of Israel, provided however, that Jordan's use will not harm the quantity or quality of the above Israeli uses. The Joint Water Committee (outlined in Article VII below) will survey existing uses for documentation and prevention of appreciable harm.

d. Jordan is entitled to an annual quantity of (10) MCM of desalinated water from the desalination of about (20) MCM of saline springs now diverted to the Jordan River. Israel will explore the possibility of financing the operation and maintenance cost of the supply to Jordan of this desalinated water (not including capital cost). Until the desalination facilities are operational, and upon the entry into force of the Treaty, Israel will supply Jordan (10) MCM of Jordan River water from the same location as in (2.a) above, outside the summer period and during dates Jordan selects, subject to the maximum capacity of transmission.

3. Additional Water

Israel and Jordan shall cooperate in finding sources for the supply to Jordan of an additional quantity of (50) MCM/year of water of drinkable standards. To this end, the Joint Water Committee will develop, within one year from the entry into force of the Treaty, a plan for the supply to Jordan of the above mentioned additional water. This plan will be forwarded to the respective governments for discussion and decision.

4. Operation and Maintenance

a. Operation and maintenance of the systems on Israeli territory that supply Jordan with water, and their electricity supply, shall be Israel's responsibility. The operation and maintenance of the new systems that serve only Jordan will be contracted at Jordan's expense to authorities or companies selected by Jordan.

b. Israel will guarantee easy unhindered access of personnel and equipment to such new systems for operation and maintenance. This subject will be further detailed in the agreements to be signed between Israel and the authorities or companies selected by Jordan.

ARTICLE II: STORAGE

1. Israel and Jordan shall cooperate to build a diversion/storage dam on the Yarmouk River directly downstream of the point 121/Adassiya Diversion. The purpose is to improve the diversion efficiency into the King Abdullah Canal of the water allocation of the Hashemite Kingdom of Jordan, and possibly for the diversion of Israel's allocation of the river water. Other purposes can be mutually agreed.

2. Israel and Jordan shall cooperate to build a system of water storage on the Jordan River, along their common boundary, between its confluence with the Yarmouk River and its confluence with Tirat Zvi/Wadi Yabis, in order to implement the provision of paragraph (2.b) of Article I above. The storage system can also be made to accommodate more floods; Israel may use up to (3) MCM/year of added storage capacity.

3. Other storage reservoirs can be discussed and agreed upon mutually.

ARTICLE III: WATER QUALITY AND PROTECTION

1. Israel and Jordan each undertake to protect, within their own jurisdiction, the shared waters of the Jordan and Yarmouk Rivers, and Arava/Araba groundwater, against any pollution, contamination, harm or unauthorized withdrawals of each other's allocations.

2. For this purpose, Israel and Jordan will jointly monitor the quality of water along their boundary, by use of jointly established monitoring stations to be operated under the guidance of the Joint Water Committee.

3. Israel and Jordan will each prohibit the disposal of municipal and industrial wastewater into the course of the Yarmouk or the Jordan Rivers before they are treated to standards allowing their unrestricted agricultural use. Implementation of this prohibition shall be completed within three years from the entry into force of the Treaty.

4. The quality of water supplied from one country to the other at any given location shall be equivalent to the quality of the water used from the same location by the supplying country.

5. Saline springs currently diverted to the Jordan River are earmarked for desalination within four years. Both countries shall cooperate to ensure that the

resulting brine will not be disposed of in the Jordan River or in any of its tributaries.

6. Israel and Jordan will each protect water systems in its own territory, supplying water to the other, against any pollution, contamination, harm or unauthorised withdrawal of each other's allocations.

ARTICLE IV: GROUNDWATER IN EMEK HA'ARAVA/WADI ARABA

1. In accordance with the provisions of this Treaty, some wells drilled and used by Israel along with their associated systems fall on the Jordanian side of the borders. These wells and systems are under Jordan's sovereignty. Israel shall retain the use of these wells and systems in the quantity and quality detailed in Appendix to this Annex, that shall be jointly prepared by 31st December, 1994. Neither country shall take, nor cause to be taken, any measure that may appreciably reduce the yields or quality of these wells and systems.

2. Throughout the period of Israel's use of these wells and systems, replacement of any well that may fail among them shall be licensed by Jordan in accordance with the laws and regulations then in effect. For this purpose, the failed well shall be treated as though it was drilled under license from the competent Jordanian authority at the time of its drilling. Israel shall supply Jordan with the log of each of the wells and the technical information about it to be kept on record. The replacement well shall be connected to the Israeli electricity and water systems.

3. Israel may increase the abstraction rate from wells and systems in Jordan by up to (10) MCM/year above the yields referred to in paragraph 1 above, subject to a determination by the Joint Water Committee that this undertaking is hydrogeologically feasible and does not harm existing Jordanian uses. Such increase is to be carried out within five years from the entry into force of the Treaty.

4. Operation and Maintenance
    a. Operation and maintenance of the wells and systems on Jordanian territory that supply Israel with water, and their electricity supply shall be Jordan's responsibility. The operation and maintenance of these wells and systems will be contracted at Israel's expense to authorities or companies selected by Israel.
    b. Jordan will guarantee easy unhindered access of personnel and equipment to such wells and systems for operation and maintenance. This subject will be further detailed in the agreements to be signed between Jordan and the authorities or companies selected by Israel.

ARTICLE V: NOTIFICATION AND AGREEMENT

1. Artificial changes in or of the course of the Jordan and Yarmouk Rivers can only be made by mutual agreement.

2. Each country undertakes to notify the other, six months ahead of time, of any intended projects which are likely to change the flow of either of the above rivers along their common boundary, or the quality of such flow. The subject

will be discussed in the Joint Water Committee with the aim of preventing harm and mitigating adverse impacts such projects may cause.

ARTICLE VI: CO-OPERATION

1. Israel and Jordan undertake to exchange relevant data on water resources through the Joint Water Committee.

2. Israel and Jordan shall co-operate in developing plans for purposes of increasing water supplies and improving water use efficiency, within the context of bilateral, regional or international cooperation.

ARTICLE VII: JOINT WATER COMMITTEE

1. For the purpose of the implementation of this Annex, the Parties will establish a Joint Water Committee comprised of three members from each country.

2. The Joint Water Committee will, with the approval of the respective governments, specify its work procedures, the frequency of its meetings, and the details of its scope of work. The Committee may invite experts and/or advisors as may be required.

3. The Committee may form, as it deems necessary, a number of specialized sub-committees and assign them technical tasks. In this context, it is agreed that these sub-committees will include a northern sub-committee and a southern sub-committee, for the management on the ground of the mutual water resources in these sectors.

The Israel-PLO Interim Agreement
28 SEPTEMBER 1995

ANNEX III: PROTOCOL CONCERNING CIVIL AFFAIRS

Article 40: Water and Sewage

On the basis of good-will, both sides have reached the following agreement in the sphere of Water and Sewage:

Principles

1. Israel recognizes the Palestinian water rights in the West Bank. These will be negotiated in the permanent status negotiations and settled in the Permanent Status Agreement relating to the various water resources.

2. Both sides recognize the necessity to develop additional water for various uses.

3. While respecting each side's powers and responsibilities in the sphere of water and sewage in their respective areas, both sides agree to coordinate the management of water and sewage resources and systems in the West Bank during the interim period, in accordance with the following principles:

a. Maintaining existing quantities of utilization from the resources, taking into consideration the quantities of additional water for the Palestinians from the Eastern Aquifer and other agreed sources in the West Bank as detailed in this Article.

b. Preventing the deterioration of water quality in water resources.

c. Using the water resources in a manner which will ensure sustainable use in the future, in quantity and quality.

d. Adjusting the utilization of the resources according to variable climatological and hydrological conditions.

e. Taking all necessary measures to prevent any harm to water resources, including those utilized by the other side.

f. Treating, reusing or properly disposing of all domestic, urban, industrial, and agricultural sewage.

g. Existing water and sewage systems shall be operated, maintained and developed in a coordinated manner, as set out in this Article.

h. Each side shall take all necessary measures to prevent any harm to the water and sewage systems in their respective areas.

i. Each side shall ensure that the provisions of this Article are applied to all resources and systems, including those privately owned or operated, in their respective areas.

Transfer of Authority

4. The Israeli side shall transfer to the Palestinian side, and the Palestinian side shall assume, powers and responsibilities in the sphere of water and sewage in the West Bank related solely to Palestinians, that are currently held by the military government and its Civil Administration, except for the issues that will be negotiated in the permanent status negotiations, in accordance with the provisions of this Article.

5. The issue of ownership of water and sewage related infrastructure in the West Bank will be addressed in the permanent status negotiations.

Additional Water

6. Both sides have agreed that the future needs of the Palestinians in the West Bank are estimated to be between 70–80 MCM/year.

7. In this framework, and in order to meet the immediate needs of the Palestinians in fresh water for domestic use, both sides recognize the necessity to make available to the Palestinians during the interim period a total quantity of 28.6 MCM/year, as detailed below:

a. Israeli Commitment:

1. Additional supply to Hebron and the Bethlehem area, including the construction of the required pipeline—1 MCM/year.

2. Additional supply to Ramallah area—0.5 MCM/year.

3. Additional supply to an agreed take-off point in the Salfit area—0.6 MCM/year.

4. Additional supply to the Nablus area—1 MCM/year.

5. The drilling of an additional well in the Jenin area—1.4 MCM/year.

6. Additional supply to the Gaza Strip—5 MCM/year.

7. The capital cost of items (1) and (5) above shall be borne by Israel.

b. Palestinian Responsibility:

1. An additional well in the Nablus area—2.1 MCM/year.

2. Additional supply to the Hebron, Bethlehem and Ramallah areas from the Eastern Aquifer or other agreed sources in the West Bank—17 MCM/year.

3. A new pipeline to convey the 5 MCM/year from the existing Israeli water system to the Gaza Strip. In the future, this quantity will come from desalination in Israel.

4. The connecting pipeline from the Salfit take-off point to Salfit.

5. The connection of the additional well in the Jenin area to the consumers.

6. The remainder of the estimated quantity of the Palestinian needs mentioned in paragraph 6 above, over the quantities mentioned in this paragraph (41.4–51.4 MCM/year), shall be developed by the Palestinians from the Eastern Aquifer and other agreed

sources in the West Bank. The Palestinians will have the right to utilize this amount for their needs (domestic and agricultural).

8. The provisions of paragraphs 6–7 above shall not prejudice the provisions of paragraph 1 to this Article.

9. Israel shall assist the Council in the implementation of the provisions of paragraph 7 above, including the following:

a. Making available all relevant data.

b. Determining the appropriate locations for the digging of wells.

10. In order to enable the implementation of paragraph 7 above, both sides shall negotiate and finalize as soon as possible a Protocol concerning the above projects, in accordance with paragraphs 18–19 below.

The Joint Water Committee

11. In order to implement their undertakings under this Article, the two sides will establish, upon the signing of this Agreement, a permanent Joint Water Committee (JWC) for the interim period, under the auspices of the CAC.

12. The function of the JWC shall be to deal with all water and sewage related issues in the West Bank including, inter alia:

a. Coordinated management of water resources.

b. Coordinated management of water and sewage systems.

c. Protection of water resources and water and sewage systems.

d. Exchange of information relating to water and sewage laws and regulations.

e. Overseeing the operation of the joint supervision and enforcement mechanism.

f. Resolution of water and sewage related disputes.

g. Cooperation in the field of water and sewage, as detailed in this Article.

h. Arrangements for water supply from one side to the other.

i. Monitoring systems. The existing regulations concerning measurement and monitoring shall remain in force until the JWC decides otherwise.

j. Other issues of mutual interest in the sphere of water and sewage.

13. The JWC shall be comprised of an equal number of representatives from each side.

14. All decisions of the JWC shall be reached by consensus, including the agenda, its procedures and other matters.

15. Detailed responsibilities and obligations of the JWC for the implementation of its functions are set out in Schedule 8.

Supervision and Enforcement Mechanism

16. Both sides recognize the necessity to establish a joint mechanism for supervision over and enforcement of their agreements in the field of water and sewage, in the West Bank.

17. For this purpose, both sides shall establish, upon the signing of this Agreement, Joint Supervision and Enforcement Teams (JSET), whose structure, role, and mode of operation is detailed in Schedule 9.

Water Purchases

18. Both sides have agreed that in the case of purchase of water by one side from the other, the purchaser shall pay the full real cost incurred by the supplier, including the cost of production at the source and the conveyance all the way to the point of delivery. Relevant provisions will be included in the Protocol referred to in paragraph 19 below.

19. The JWC will develop a Protocol relating to all aspects of the supply of water from one side to the other, including, inter alia, reliability of supply, quality of supplied water, schedule of delivery and off-set of debts.

Mutual Cooperation

20. Both sides will cooperate in the field of water and sewage, including, inter alia:

a. Cooperation in the framework of the Israeli-Palestinian Continuing Committee for Economic Cooperation, in accordance with the provisions of Article XI and Annex III of the Declaration of Principles.

b. Cooperation concerning regional development programs, in accordance with the provisions of Article XI and Annex IV of the Declaration of Principles.

c. Cooperation, within the framework of the joint Israeli-Palestinian-American Committee, on water production and development related projects agreed upon by the JWC.

d. Cooperation in the promotion and development of other agreed water-related and sewage-related joint projects, in existing or future multi-lateral forums.

e. Cooperation in water-related technology transfer, research and development, training, and setting of standards.

f. Cooperation in the development of mechanisms for dealing with water-related and sewage-related natural and man-made emergencies and extreme conditions.

g. Cooperation in the exchange of available relevant water and sewage data, including:

1. Measurements and maps related to water resources and uses.

2. Reports, plans, studies, researches and project documents related to water and sewage.

3. Data concerning the existing extractions, utilization and estimated potential of the Eastern, North-Eastern and Western Aquifers (attached as Schedule 10).

Protection of Water Resources and Water and Sewage Systems

21. Each side shall take all necessary measures to prevent any harm, pollution, or deterioration of water quality of the water resources.

22. Each side shall take all necessary measures for the physical protection of the water and sewage systems in their respective areas.

23. Each side shall take all necessary measures to prevent any pollution or contamination of the water and sewage systems, including those of the other side.

24. Each side shall reimburse the other for any unauthorized use of or sabotage to water and sewage systems situated in the areas under its responsibility which serve the other side.

The Gaza Strip

25. The existing agreements and arrangements between the sides concerning water resources and water and sewage systems in the Gaza Strip shall remain unchanged, as detailed in Schedule 11.

Acidity is the condition of water or soil which contains a sufficient amount of acid substances to lower the pH below 7.0. pH is an expression of the intensity of the basic or acid condition of a liquid. The pH may range from 0 to 14, with 7 representing neutrality; numbers less than 7 indicate increasing acidity and numbers greater than 7 indicate increasing alkalinity. Natural waters usually have a pH between 6.5 and 8.5.

Agricultural Water Use usually includes livestock, animal specialty, and irrigation water use.

Alawi is an offshoot of Shia Islam. Alawis are minorities that live primarily in Syria and eastern Turkey. The president of Syria, Hafez al-Assad, is an Alawi.

Aquifer see groundwater

Arid is an adjective applied to regions where precipitation is deficient in quantity, thus making agriculture impractical without irrigation. Lands that receive less than 250 mm of precipitation per year are usually considered unfit or unprofitable for rain-fed farming. The area south of the Fertile Crescent is considered arid.

Balfour Declaration (1917) states that the British government "views with favour" the creation of a national home for Jews in Palestine.

Ba'th (Arab Socialist Resurrection Party) is a pan-Arab party advocating unity, freedom, and socialism for the Arabs. The party became prominent in the 1950s in Syria and Iraq. The two countries have been run by rival Ba'ath governments since the 1970s.

Camp David Accord was a peace treaty that ended decades of war between Israel and Egypt. It resulted in Israel's return of the occupied Sinai Desert to Egypt.

Consumptive Use of water is the quantity of water removed from the immediate water environment and not readily available for reuse—the amount of water absorbed by a crop and transpired, or used directly in the building of plant tissue, together with the water evaporated from the cropped area or the normal loss of water from the soil by evaporation and plant transpiration. It is also the quantity of water discharged to the atmosphere or incorporated in the products of vegetative growth, food processing, or an industrial process. Also referred to as water consumed.

Conveyance Loss is when water is lost in transit through leakage or evaporation. Lost water is typically not available for further use.

Critical Groundwater Area is an area that has or may develop certain groundwater problems, such as declining water levels. There are usually limits put on the amount of development, type of land use, and volume of water withdrawn from such areas.

Demand Management is policies and schemes that induce consumers to ration their consumption, and to adopt efficient water-use practices based on the relative value of the water. Public information programs, regulatory controls and enforcement, water-conserving technology, cost recovery for the water delivery operations and their maintenance, and a fair return on capital investment are usually effective in limiting the need for increased water supplies.

Desalination is the process of removing salts from water, usually to produce potable quality water. The techniques used include distillation, electrodialysis, and reverse osmosis. The cost of desalination drops along with the level of salts in the water.

Diversion is the removal of water from a water body for use elsewhere. Diversions may be used to protect valley floors or buildings from hillside runoff.

Divide see watershed

Domestic Consumption is water used for normal everyday household purposes, including watering lawns and gardens. Water used for domestic consumption may be supplied by public and/or private enterprises. This is also called residential or municipal water use.

Drainage Basin is the area of land that drains surface water from precipitation, as well as sediment and dissolved materials, to a common outlet at some point along a stream channel.

Drip Irrigation is a method of microirrigation wherein small quantities of water are frequently applied to the soil surface as drops or small streams through emitters.

Drought may be defined as less than average precipitation over a certain period of time, or less than what people "need" or have planned for.

Druze (or Druse) is an offshoot of Islam that has developed its own rituals and practices and a close-knit community structure. Small numbers of Druze live in Lebanon, Jordan, Syria, and Israel.

Fresh Water is water that contains less than 1,000 mg/L of dissolved solids. Generally, more than 500 mg/L of dissolved solids is undesirable for drinking and many industrial uses.

Groundwater includes all subsurface water in the saturated zone (a zone in which all voids are filled with water). It often supplies wells and springs. Because groundwater is a major source of drinking water in the Jordan River basin states, there is growing concern over critical areas where leaching agricultural or industrial pollutants or substances are contaminating groundwater. Groundwater may not be recharged (fossil water), or it may be artificially or naturally recharged.

Hamas is an Arabic acronym for the Islamic Resistance Movement. It is a multifaceted Islamic organization which originated in Gaza. There are social as well as military dimensions to its activities. It opposes the Oslo Accord. It is considered an illegal party in Israel.

Hydrological Cycle (water cycle) is the movement of water from the atmosphere to the earth and back to the atmosphere through various processes. These processes include: precipitation, infiltration, percolation, storage, evaporation, transpiration, and condensation.

Hydrology is the study of water movement on the Earth's surface and in underlying soils and rocks.

In-Channel Use see instream use

Industrial Water Use is the use of water for a variety of industrial purposes, ranging from fabrication to cooling and petroleum refining. Water use here may be consumptive or nonconsumptive. The water may be obtained from public or private suppliers.

Instream Use is water use that can be carried out without removing or diverting the water from its ground- or surface-water source. This use includes navigation, water-quality improvement, fish and wildlife propagation, and recreation. Hydroelectric water use is often classified as an instream use.

Intermittent (or seasonal) *Stream* is one that flows periodically when it receives water from springs, rainfall, or surface sources such as melting snow.

Intifadah (Arabic for "shaking off") is a grassroots Palestinian uprising against Israeli occupation. It was initiated in 1987 by Palestinians inside Gaza and the West Bank, and ended in 1994 following the signing of the Oslo Peace Accord between the Palestinians and the State of Israel.

Irrigation is the controlled application of water for cultural purposes through artificial systems to supply water requirements not satisfied by rainfall.

Kibbutz is a collective settlement in Israel where land and property are communally owned. Kibbutzes have been increasingly relying on small industries to supplement their traditional reliance on agriculture.

Labour is a left-of-center political party in Israel. In the 1980s and till the mid-1990s, it was led by Yitzhak Rabin, Shimon Peres, and Ehud Barak.

Land-For-Peace is a shorthand description of the United Nations Security Council Resolution 242, which called on Israel to return lands it had occupied during the 1967 war in return for peaceful relations with its neighbors.

Likud is a right-of-center political party in Israel. In the 1980s and till the mid-1990s, Likud was led by Menachem Begin, Yitzhak Shamir, and Benjamin Netanyahu.

Muhafaza is the Arabic word for province. The term is used in Syria and Lebanon. Provinces (muhafazat) are subdivided into aqdiya (cazas), or subdistricts.

Natural Flow is the rate of water movement past a specified point on a natural stream whose flow is uninterrupted by storage, import, export, return flow, or change in consumptive use caused by human-controlled modifications to land use.

Natural Resource is any element, material, or organism existing in nature that may be useful to humans.

Nonconsumptive use is the human use of water in a way that does not decrease its supply. Fishing and bathing are examples of this.

Non-Point Pollution is when there are spatially diffuse sources of pollution. Pollutants are introduced into a receiving stream from unspecific outlets. Generally, the pollutants originate from land-use activities, and are carried to lakes and streams by surface runoff. The commonly used categories for non-point sources are: agriculture, forestry, urban, mining, construction, dams and channels, land disposal, and saltwater intrusion. Compare point source.

Offstream Use is water withdrawn or diverted from a ground- or surface-water source. It is sometimes called off-channel use or withdrawal.

Orographic Precipitation is the rainfall that occurs as a result of warm, humid air being forced to rise by topographic features such as mountains. That precipitation levels on the western slopes of the mountain range paralleling the coastal plain of Syria, Lebanon, and northeastern Israel, and on the hills of the West Bank, are significantly higher than those in the interior is partially explained by the orographic effect.

Oslo Peace Accord, or the Declaration of Principles, was a document signed in 1993 by the PLO, Israel, and the United States, according to which the antagonists agreed to negotiate an end to their territorial claims, based on UN Security Council Resolution 242 (the so-called "land-for-peace" formula). It committed Israel to returning occupied Palestinian territories in return for peace.

Participation is a process in which stakeholders (the public, government, and planners) influence policy formulation, alternative designs, investment choices, and management decisions affecting their communities and establish the necessary sense of ownership. Participation can help coordinate interests, increase transparency and accountability in decision-making, and encourage user ownership—all of which increase the probability of a project's success.

People Overpopulation is when there are too many people in a given geographical area. Even if those people use few (water) resources per person (the minimum amount they need to survive), overpopulation results in pollution, environmental degradation, and resource depletion.

Perennial Stream is one that flows all year round.

PLO (Palestine Liberation Organization). Established in 1964 to develop political and military strategies for the creation of a sovereign Palestinian state in the pre-1967 Israel. Overall PLO authority is vested in the Palestine National Council (PNC). In 1988, the PLO recognized the State of Israel and revised its territorial objectives, endorsing a two-state solution, with a Palestinian one on the West Bank and Gaza Strip.

Point Source Pollution is water pollution that results from the discharges into receiving waters from a stationary location or fixed facility. It is an easily identifiable source of pollution. Common point sources of pollution are discharges from factories and municipal sewage treatment plants. Compare to non-point source.

ppb (parts per billion): number of parts of a chemical found in one billion parts of a solid, liquid, or gaseous mixture. Equivalent to micrograms per liter.

ppm (parts per million): number of parts of a chemical found in one million parts of a solid, liquid, or gaseous mixture. Equivalent to milligrams per liter.

Precipitation is the hail, rain, mist, sleet, and snow that are deposited on Earth.

Recharge is a process by which rainwater (or precipitation) seeps for sometimes thousands of meters into aquifers through highly porous soil, faults, or fractures.

Reclaimed Water is wastewater that has been treated and diverted for reuse before it reaches a natural waterway or aquifer.

Riparian Water Right is the legal right, held by an owner of land contiguous to or bordering on a natural stream or lake, to take water from the source for use on the contiguous land. This right may include the right to prevent diversion or misuse of upstream waters. Riparian land is land that borders on surface water.

Runoff is the movement of fresh water from precipitation and snowmelt to rivers, lakes, wetlands, and ultimately, the ocean. Runoff is thought of as a lost resource and a contributor to non-point source pollution.

Saline Water contains more than 1,000 parts per million (ppm) (or 1,000 milligrams per liter) of dissolved solids.

Shia (also Shiite, Shi'i) means party or partisans of Ali, Prophet Muhammad's cousin. It refers to those Muslims who supported Ali ibn Abi Talib as Muhammad's rightful and designated successor (Caliph). It is one of the two major sects in Islam. Shia are a minority within the Middle East, except in Iran, where they constitute the overwhelming majority. There are significant Shia minorities in Lebanon, Bahrain, Saudi Arabia, Kuwait, and Iraq.

Six-Day War (also June 1967 War) was preceded by political and military tensions between Israel and its neighbors. Israel attacked Egypt, Jordan, and Syria, capturing the Sinai Peninsula, West Bank, Gaza Strip, and Golan Heights.

Streamflow is the discharge that occurs in a natural channel. The term may be applied to discharge whether or not it is affected by diversion or regulation.

Sunni refers to the custom or procedure, the code of acceptable behavior, for Muslims based on the Koran and hadith. It is the largest sect of Islam in the Middle East and throughout the Muslim world, except for Iran.

Suq (Souk) is (1) a public weekly market in rural areas in the Middle East. It is usually held in the same village on the same day of the week, so that the village may have the word incorporated into its name. It also refers to (2) a section of a Middle Eastern city that is devoted to the wares and work of potters, cloth merchants, wood workers, spice sellers, etc.

Surface Irrigation is a broad class of wasteful irrigation methods in which water is distributed over the soil surface by gravity flow.

Surface Water is water in lakes, bays, ponds, impounding reservoirs, springs, rivers, streams, creeks, estuaries, wetlands, marshes, inlets, canals, gulfs inside the territorial limits of the state, and all other bodies of surface water, natural or artificial, inland or coastal, fresh or salt, navigable or nonnavigable, and including the beds and banks of all watercourses and bodies of surface water, that are wholly or partially inside or bordering the country.

Sustainable Water Management is exploitation of water resources in such a way that it can be carried on indefinitely. Removal of water from an aquifer in excess of recharge is, in the long term, not a sustainable management method.

UNHCR (United Nations High Commission for Refugees) was established in 1951 to provide international protection and material assistance to refugees worldwide. UNHCR has several refugee projects in the Middle East.

UNIFIL are the United Nations Interim Forces in Lebanon. They were formed and deployed after Israel's 1978 "Litani Operation" and occupation of southern Lebanon up to the Litani River.

UNRWA is the United Nations Refugee and Works Agency.

UN Security Council Resolution 242 calls on Israel to "withdraw from territories" occupied during the 1967 Six-Day War.

Watershed is the land area from which water drains toward a common watercourse in a natural basin. An area of land that contributes runoff to one specific delivery point; large watersheds may be composed of several smaller "subsheds," each of which contributes runoff to different locations that ultimately combine at a common delivery point.

Water Table is the level below the Earth's surface at which the ground becomes saturated with water. The surface of an unconfined aquifer which fluctuates due to seasonal precipitation. This level can be very near the surface of the ground or far below it.

Geographical Terms

Bahr (sing.), Bahrain (pl.)	sea, lake
Beit	house
Bekaa or Biqa	fertile plain
Bir	well
Birkeh, Birket	pool, tank
Dar	dwelling
Deir	monastery
El (al, em, en, er, esh, et)	the
Gezira, Jezireh	island
Ghor	hollow, valley
Ibn	son of
Jabal	hill, mountain
Kabir	great, large
Kefr	village
Khan	night stopping place for caravans
Mar	saint
Merj, Marj	plain
Nahr	river
Qanat (P)	canal
Ramle (h)	sand
Ras	cape, headland
Sahel	plain, coastline
Tel, Tell (pl. Tulul)	small hill

Tony Allan is Professor of Geography at the School of Oriental and African Studies at the University of London (UK). He specializes in the analysis of water resources in arid and semiarid regions, especially the surface, soil, and ground waters of the Middle East and North Africa. He has published works on the Nile and the Jordan basins and on the region as a whole. His recent research has emphasized the rich explanation which comes from the approach of political economy and especially from social theory. He has demonstrated that it is social and economic policies which will ensure the long-term "water security" of water-deficit regions such as the Middle East and North Africa. Professor Allan also contributes to studies and advisory groups of the World Bank and of other international agencies and governments.

Peter Beaumont has been Professor of Geography at the University of Wales, Lampeter, since 1978. Before this he was a Lecturer and Senior Lecturer in the Department of Geography at the University of Durham. He has been a Harkness Fellow at the University of California, Berkeley (1969–1971); a NATO Fellow (1972–1973); NATO Fellow (Committee on the Challenges of Modern Society) (1990–1992); Visiting Fellow and Scholar, Wolfson College, Oxford (1989 and 1998); Visiting Fellow at the Oxford Center for Islamic Studies, Oxford (1989 and 1998); Visiting Professor at the University of California, Berkeley (1971) and at the University of Texas, Austin (1983). His major research work has focused on water resource and environmental management issues in the Middle East and other dryland regions.

Steve Lonergan (BSc, Duke University; PhD, University of Pennsylvania). Professor, Department of Geography, University of Victoria. Dr. Lonergan has been on the faculty of McMaster University and was a Visiting Fellow at the University of Auckland. He is past Director of the Centre for Regional Sustainable Development at the University of Victoria and presently directs the project on Global Environmental Change and Human Security for the International Human Dimensions Programme on Global Environmental Change. He has authored numerous articles on Middle East water issues, and is coauthor with David Brooks of *Watershed: The Role of Fresh Water in the Israeli-Palestinian Conflict* (IDRC Books, 1994). Dr. Lonergan is presently writing a book on water resource development on the Euphrates River.

Aaron T. Wolf (MS, PhD, University of Wisconsin, Madison) is an Assistant Professor of Geography in the Department of Geosciences, Oregon State University, whose research and teaching focus on the interaction between water science and water policy, particularly as related to conflict and conflict resolution. He has acted as consultant to the U.S. Department of State, the U.S. Agency for International Development, and the World Bank on various aspects of international water resources and dispute resolution. He was involved in developing the strategies for resolving water aspects of the Arab-Israeli conflict, including coauthoring a State Department reference text, and participating in both official and "Track II" meetings among co-riparians. He is author of *Hydropolitics along the Jordan River: Scarce Water and Its Impact on the Arab-Israeli Conflict* (United Nations University Press, 1995) and a coauthor of *Core and Periphery: A Comprehensive Approach to Middle Eastern Water* (Oxford University Press, 1997).

Hussein A. Amery (BA, University of Calgary; MA, Wilfrid Laurier University; PhD, McMaster University) is an assistant professor at the Colorado School of Mines. Before leaving Canada, he worked in the geography departments of the University of Toronto, Bishop's University, and the University of Lethbridge. His research interests and publications are in the area of environmental security in the context of water scarcity in the Middle East, with particular interest in the Litani River and the Jordan River basin states. He is also published in the area of migrants' remittances and their consequences on rural economic development in Lebanon. He was a consultant on water-related issues for development and research institutions, including the Center for International Studies (University of Toronto) and International Development and Research Center (Ottawa).

Frederic C. Hof is a partner in the international business development firm Armitage Associates, a corporate member of the World Water Council. Prior to the founding of Armitage Associates in 1993, Mr. Hof served in several capacities in the U.S. Departments of Defense and State, including U.S. Army Attaché in Beirut, Lebanon (1981–1982), and senior member of the U.S. team mediating the Israel-Jordan Unity Dam dispute

(1989–1990). He is a member of the National Advisory Committee of the Middle East Policy Council.

The author of *Galilee Divided* (Westview Press, 1985), Mr. Hof has written extensively on Middle Eastern water matters in *Middle East Policy, The Journal of Commerce,* and elsewhere. His interests in the water field focus on management of transboundary resources and dispute resolution.

Paul Kay has a BSc (1971) from the University of Toronto, and an MS (1973) and a PhD (1976) from the University of Wisconsin, Madison. He has held regular or visiting faculty positions at the University of Utah, Clark University, Haifa University, and the University of Nebraska, Lincoln. He is currently Associate Professor of Environment and Resource Studies at the University of Waterloo. He specializes in implications of climatic variability for water resources.

Dr. Kay has studied climatic variations and their implications on a variety of time scales, from Quaternary to modern, and in a variety of settings, including northern Canada, the U.S. intermountain west, and the Middle East. He has published numerous papers and conference reports on issues of climate, variation, and the management of water in the Israeli setting.

Bruce Mitchell has a BA (1966) and an MA (1967) from the University of British Columbia, and a PhD (1969) from the University of Liverpool. He has been Chairman of Geography and Associate Dean of Graduate Studies and Research for the Faculty of Environmental Studies, and currently is Professor of Geography, at the University of Waterloo. He specializes in institutional and policy aspects of water and integrated resource management. Professor Mitchell has studied water management for over thirty years, and has written or edited twenty-one books and over a hundred articles.

In 1992, he was one of three co-organizers of the Conference on the Middle East Water Crisis: Creative Perspectives and Solutions, which was supported by the Canadian International Development Agency, the Canadian Institute for International Peace and Security, and the National Council on Canada-Arabic Relations. Selected papers from the conference were published in a theme issue of *Water International* in 1993. He is presently Program Director for an environmental capacity building project in Sulawesi, Indonesia. He is a Past President of the Canadian Water Resources Association.

Nurit Kliot is a professor in the Department of Geography, University of Haifa. Her teaching and research focus on political geography, environmental policy, and water resources. She is the author of *Water Re-sources and Conflict in the Middle East* (Routledge, 1994) and (with D. Shmueli and U. Shamir) *Institutional Frameworks for the Manage-ment of Transboundary Water Resources* (2 vols.) (Technion Water Research Institute, 1997).

Gwyn Rowley is Reader in Geography at the University of Sheffield and a Peabody Fellow at Harvard University. He has longtime interests in the Middle East in general and the Mashreq in particular, where his research centers on competitions over space in and about the Occupied Territories adjacent to Israel. He places prime importance upon personal face-to-face fieldwork in his continuing research programs.

John Kolars is Professor Emeritus (1993), the University of Michigan, Ann Arbor, where he taught Geography and Middle Eastern Studies (1963–1994). He received his BSc (Geology) and MA (Geography) at the University of Washington. His research has focused upon agricultural development and water use in rural Turkey and the basins of the Euphrates, Tigris, and Litani (Lebanon) Rivers. He is a senior consultant with Associates for Middle Eastern Research and the Middle East Water Commission of the International Water Resources Association, and a regular lecturer on the ecology of the Middle East at the Department of State, National Foreign Affairs Training Center, where he has been awarded the title of Distinguished Visiting Lecturer. He frequently participates in "second track" mediation (meetings) between water disputants in the Middle East. He has published numerous works on water in the Middle East, including a definitive, book-length study of the Southeast Anatolia Development Project (Turkish acronym GAP) and the Euphrates-Tigris Rivers (coauthored with Col. William A. Mitchell), a monograph on the character and usage of the Litani River, and numerous articles on the impact of such developments on both the former two rivers and the gulf into which they flow. He has been a consultant on Middle East water problems to the U.S. Department of State, U.S. Agency for International Development, the U.S. Army Corps of Engineers, the Industrial College of the Armed Forces of the NDU, the White House Council on Environmental Quality, Battelle PNNL, and the Kingdom of Jordan, among numerous others.

INDEX